Current and Future Developments in Physiology
(Volume 1)
Control of Pancreatic Beta Cell Function and Plasticity in Health and Diabetes

Edited by

David J. Hill

Lawson Health Research Institute
St. Joseph's Health Care,
268 Grosvenor Street, London,
Ontario N6A 4V2,
Canada

advertisements or ideas contained in the Work.

Limitation of Liability:

In no event will Bentham Science Publishers, its staff, editors and/or authors, be liable for any damages, including, without limitation, special, incidental and/or consequential damages and/or damages for lost data and/or profits arising out of (whether directly or indirectly) the use or inability to use the Work. The entire liability of Bentham Science Publishers shall be limited to the amount actually paid by you for the Work.

General:

1. Any dispute or claim arising out of or in connection with this License Agreement or the Work (including non-contractual disputes or claims) will be governed by and construed in accordance with the laws of the U.A.E. as applied in the Emirate of Dubai. Each party agrees that the courts of the Emirate of Dubai shall have exclusive jurisdiction to settle any dispute or claim arising out of or in connection with this License Agreement or the Work (including non-contractual disputes or claims).
2. Your rights under this License Agreement will automatically terminate without notice and without the need for a court order if at any point you breach any terms of this License Agreement. In no event will any delay or failure by Bentham Science Publishers in enforcing your compliance with this License Agreement constitute a waiver of any of its rights.
3. You acknowledge that you have read this License Agreement, and agree to be bound by its terms and conditions. To the extent that any other terms and conditions presented on any website of Bentham Science Publishers conflict with, or are inconsistent with, the terms and conditions set out in this License Agreement, you acknowledge that the terms and conditions set out in this License Agreement shall prevail.

Bentham Science Publishers Ltd.
Executive Suite Y - 2
PO Box 7917, Saif Zone
Sharjah, U.A.E.
Email: subscriptions@benthamscience.org

**BENTHAM
SCIENCE**

CONTENTS

FOREWORD

The incidence of both type 1 (T1D) and type 2 diabetes (T2D) has increased over the past few decades and is frequently described as an epidemic. It is now recognized that loss of β-cell mass occurs in both forms of diabetes, through autoimmune destruction in the case of T1D and by exhaustion in the case of T2D. Despite current treatments, blood glucose levels are not restored to normal and this leads to serious injury in several organ systems. Ideally it should be possible to treat or reverse both forms of diabetes by re-establishing sufficient functioning β-cells to maintain normoglycemia. There is general agreement that the cells in the adult pancreas can regenerate, but the routes by which this occurs remain controversial. New β-cells are produced early in life by replication of existing cells, but it is less clear how β-cells could be produced naturally or induced in adults through post-injury ductal neogenesis, activation of resident progenitors/stem cells or transdifferentiation of other non-β-cells (*e.g.*, α-cells or acinar cells). A further complication is understanding to what extent these processes are influenced by environmental conditions.

It has recently been appreciated that terminally differentiated and stem/progenitor cells can differentiate into other cells in the pancreas, consistent with the idea of plasticity in cell fate that depends in part on modification of developmental programs by environmental factors [1]. The fate of cells depends on the type of stress encountered, for example pregnancy, metabolic demand due to hyperglycemia and insulin resistance (glucotoxicity), or ectopic expression or deletion of lineage specific factors.

The importance of restoring lost β-cell mass in T1D has been demonstrated by transplanting islets from cadaveric donors. Unfortunately, this solution is only temporary, requires the recipient to be on powerful immunosuppressive drugs and has side effects. Further constraints are the requirement for 2-3 pancreata/recipient and the limited number of pancreata available for transplantation. Hence, a strong research effort is underway to find other sources of β-cells. These include transplanting pig islets, promoting transdifferentiation of non-β-cells into functioning, insulin-producing β-like cells or enhancing proliferation, maintenance and function of remaining β-cells. Several attempts have been made to develop functioning β-cells from embryonic or adult pluripotent stem cells. However, most reports described the production of immature β-like cells from human pluripotent stem cells that did not respond normally to glucose. A recent report from the Melton group described the *in vitro* production from human embryonic stem cells of glucose-sensitive, insulin-producing β-cells that resemble mature β-cells [2]. It has been pointed out that the basis of this (and other) attempts to find new sources of β-cells comes from a vastly improved and expanded knowledge of pancreas developmental biology [3] and the conditions required to change cell fate, areas which are addressed in this e-book. As with any discovery, there is good news and bad news. The use of embryonic stem cells still raises ethical concerns and there are technical issues to overcome before such a therapy can be considered for use in the clinic. Others must replicate these findings. It must also be remembered that these cells do not have the exact genetic profile and they are not identical copies of β-cells *in situ* [3]. It is unclear what will be the long term effects of transplanting such cells in patients. As with any "replacement" β-cells, they must be protected from the host's immune system. To do this, they will likely be placed in protective capsules that are themselves the target of a fibrotic process in the body that blocks the exchange of nutrients and insulin across the membrane. In short, it will be a while before these cells are ready for prime time.

Many questions remain unanswered regarding β-cell plasticity. Is it only a subset of cells of a certain lineage (*e.g.*, acinar) that respond to external stress and express transcription factors that promote self renewal or multi-potency directly or through de-differentiation? An ever present danger is the potential for these processes to go unchecked if fate constraints are lost, leading to cancer. It has been pointed out that it is difficult to evaluate transdifferentiation *per se* because the characteristics that reflect the extent of the new cell's maturity and stability of the phenotype are unclear [1]. Does the phenotype of these new cells differ from that of the cells that occur during development?

The editor, Dr. David Hill, has brought together experts in the field who provide extensive and sometimes provocative state-of-the-art discussions of various key aspects of β-cell function and plasticity in health and diabetes. The book is divided into three sections. In the first section entitled: "Origins and developmental biology of β-cells", Dr. Bertrand Duvillié and colleagues (Paris) review advances in understanding the hierarchy of transcription factors and discuss the influence of growth factors, partial oxygen pressure and nutrients in the intrauterine environment on β-cell maturation. This is followed by Aaron Cox's (Houston) discussion of the role of aging on β-cell development and how the decline in β-cell proliferation is affected by age-related impairment in signal transduction, altered cell cycle progression and epigenetic regulation of genes is likely to affect attempts to regenerate β-cells. Part 1 ends with a chapter on β-cell mass across the spectrum of aging, health and diabetes by Dr. Manami Hara and colleagues (Chicago). They discuss the heterogeneous distribution and function of β-cells among and within individuals and the need to identify markers of dysfunctional β-cells.

The second section addresses "Factors controlling β-cell mass and function". Beginning with Dr. David Hill's (London, ON) discussion of gestational programming of β-cell mass and function in which he emphasizes that the period of gestation affects cell fate and β-cell function in the offspring. Using fetal growth retardation as a model, he notes several systems are adversely affected leading to decreased β-cell mass, proliferation, increased apoptosis and later impaired glucose regulation. The mechanisms are discussed and the point is made that giving micronutrient supplements to the dam and β-cell trophic peptide hormones to the neonate decreases disease risk. The second article by Drs. Mulchand S. Patel and Saleh Mahmood (Buffalo) addresses modification of β-cell function by diet in neonates. They suggest that the feeding of carbohydrate rich food adversely affects islet structure and β cell function in suckling neonates leading to hyperinsulinemia attributable to increased β cell plasticity. The changes noted in developmental gene expression and β-cell function lead to obesity in the adult which could promote development of T2D. The third chapter in this section is by Dr. Brian T. Layden and colleagues (Chicago) in which they provide a discussion of intra-islet control of β-cell function and mass. They highlight the role of under-appreciated autocrine and paracrine factors that contribute to β-cell mass and function. The last chapter in this section by Dr. Jens HøiriisNielson and colleagues (Copenhagen) deals with beta cell adaptability during pregnancy. Although it is well known that β-cell mass expands during pregnancy, they describe in detail the many growth factors, transcriptional modifications and changes in gene expression that have recently been characterized. The mechanism involved in β-cell mass expansion during pregnancy remains unclear, in particular the degree to which neogenesis is involved in rodents and humans.

The third and last part of the book deals with "Generation of β-cells and future applications". Chapter 8 by Dr. Christine A. Beamish (London, ON) addresses the production of β-cells from embryonic and adult stem cells and progenitors as a source of replacement β-cells in diabetes. The plasticity of cell fate in the endocrine pancreas is discussed. The last chapter is by Dr. Tyler T. Cooper and colleagues (London, ON) in which they describe the latest thinking on the role of bone marrow-derived stem cells for β-cell regeneration.

It is now clear that T1D is far more complex than previously appreciated [4]. In particular, the long history of mostly immune-based therapies has not resulted in new treatments or cures. Despite more than 20 clinical trials, most focused on immune suppression, there has been a remarkable lack of success with respect to prevention or reversal of T1D [5]. The picture continues to evolve with respect to T2D as well, as a role for adipose tissue inflammation is now being investigated. To address this level of complexity, both forms of diabetes must be thought of in terms of their integrative biology. That is, how do genetic predisposition, inappropriate immune reactivity, inappropriate β-cell mass/function/regenerative capacity and environmental exposures come together to result in diabetes? We are entering a new era of big data obtained through readily available high throughput analyses and higher computer power. This has raised the possibility of personalized or so-called "precision medicine" to address the heterogeneity in the human population and reveal the many pathways by which diabetes likely occurs. A further new approach involves the acknowledgement that there is unlikely to be a single infectious agent akin to *helicobacter pylori* and ulcers in the environment that causes diabetes. To address the multiplicity of environmental factors that undoubtedly influence diabetes and other chronic diseases, a new "omics" designation was coined, termed the exposome [6]. And finally, it must be said that timing is indeed everything. Thus, the timing of these interactions and exposures is crucial in determining the success of the early developmental program for appropriate β-cell mass/function or its potential recapitulation following injury, inflammation or disease. The articles in this e-book provide up-to-date summaries of key elements affecting natural and induced forms of β-cell plasticity and its potential as a therapy for diabetes. The reader will no doubt enjoy this timely collection.

Dr. Fraser W. Scott
The Ottawa Hospital Research Institute
Department of Medicine
Department of Biochemistry, Microbiology and Immunology
University of Ottawa
Ottawa, Ontario, Canada

REFERENCES

[1] Puri S, Folias AE, Hebrok M. Plasticity and dedifferentiation within the pancreas: development, homeostasis, and disease. Cell Stem Cell 2015; 16(1): 18-31.
[http://dx.doi.org/10.1016/j.stem.2014.11.001] [PMID: 25465113]

[2] Pagliuca FW, Millman JR, Gürtler M, *et al.* Generation of functional human pancreatic β cells *in vitro*. Cell 2014; 159(2): 428-39.
[http://dx.doi.org/10.1016/j.cell.2014.09.040] [PMID: 25303535]

[3] Mfopou JK, Bouwens L. Diabetes: β cells at last. Nat Rev Endocrinol 2015; 11(1): 5-6.
[http://dx.doi.org/10.1038/nrendo.2014.200] [PMID: 25385037]

[4] Atkinson MA, von Herrath M, Powers AC, Clare-Salzler M. Current concepts on the pathogenesis of type 1 diabetesconsiderations for attempts to prevent and reverse the disease. Diabetes Care 2015; 38(6): 979-88.
[http://dx.doi.org/10.2337/dc15-0144] [PMID: 25998290]

[5] Lord S, Greenbaum CJ. Disease modifying therapies in type 1 diabetes: Where have we been, and where are we going? Pharmacol Res 2015; 98: 3-8.
[http://dx.doi.org/10.1016/j.phrs.2015.02.002] [PMID: 25771310]

[6] Wild CP. Complementing the genome with an exposome: the outstanding challenge of environmental exposure measurement in molecular epidemiology. Cancer Epidemiol Biomarkers Prev 2005; 14(8): 1847-50.
[http://dx.doi.org/10.1158/1055-9965.EPI-05-0456] [PMID: 16103423]

PREFACE

An adaptive metabolic axis has been a major evolutionary advantage in allowing humans to colonize every part of the globe from arid deserts to permanent ice fields. Prior to an effective food supply chain, metabolic plasticity evolved to deal with seasonal famines balanced by times of plenty, and a greater diversity of diets than perhaps any other mammalian species. In the developed world there are new challenges to metabolic plasticity including food over-abundance, unbalanced diets, child and adult obesity, and an increasing rate of type 1 and 2 diabetes. A plasticity of pancreatic β-cell mass and function are key to metabolic adaption. The β-cell mass normally increases proportionally to fetal and child growth, in response to the added metabolic stress of pregnancy, and in response to the nutritional stress of an obesogenic diet. Yet, in the face of the autoimmune challenge of type 1 diabetes or the glucotoxicity of type 2 diabetes there is a net loss of β-cells with limited potential for endogenous regeneration. Thus lies the paradox. How can a highly physiologically-adaptive β-cell mass prove so difficult to manipulate following the pathological loss that accompanies diabetes?

Key to creating and testing strategies for the therapeutic manipulation of β-cell number is to know their developmental origins and normal ontogeny. The first section of this volume addresses current knowledge around the developmental origins of pancreatic β-cells, and how β-cell mass and proliferation change throughout the human lifespan. The second section explores the mechanisms responsible for β-cell plasticity, drawing from animal models and clinical studies revealing environmental, epigenetic, endocrine and paracrine regulators that contribute to the normal homeostatic processes, and the delicate balance of proliferation *vs.* apoptotic loss that optimizes β-cell mass during normal metabolic homeostasis. The final section examines the presence and potential of resident stem cells within the pancreas or bone marrow, β-cell progenitors, and the potential for pancreatic endocrine cell differentiation or trans-differentiation.

Underlying each of these chapters is the assumption that β-cells can potentially be replaced endogenously, but only through a thorough understanding of normal development and the exploitation of existing, but perhaps sub-optimal adaptive physiology. There is great reason for confidence. In humans there is reproducible histological evidence of β-cell turnover involving mitogenesis and apoptosis throughout life, including both children and adults with type 1 or type 2 diabetes [1 - 5]. The regenerative potential of human β-cells may normally be age-limited, since new cells were not generated in the short-term in patients aged over 50 following surgical reduction of pancreatic mass [6]. However, insulin release had improved between 2-4 years post-surgery, suggesting that even in older individuals a slower adaptive replacement of β-cells can occur [7]. Overcoming such physiological limitations to optimize β-cell mass and function to match metabolic demand will likely be a major focus for diabetes research in the coming decade.

Dr. David J. Hill
Lawson Health Research Institute
St. Joseph's Health Care, 268 Grosvenor Street
London, Ontario N6A 4V2
Canada
Tel: 519 6466100 Ext. 64716
E-mail: david.hill@lhrionhealth.ca

REFERENCES

[1] Gepts W, De Mey J. Islet cell survival determined by morphology. An immunocytochemical study of the islets of Langerhans in juvenile diabetes mellitus. Diabetes 1978; 27 (Suppl. 1): 251-61.
[http://dx.doi.org/10.2337/diab.27.1.S251] [PMID: 75815]

[2] Yoneda S, Uno S, Iwahashi H, *et al.* Predominance of β-cell neogenesis rather than replication in humans with an impaired glucose tolerance and newly diagnosed diabetes. J Clin Endocrinol Metab 2013; 98(5): 2053-61.
[http://dx.doi.org/10.1210/jc.2012-3832] [PMID: 23539729]

[3] Keenan HA, Sun JK, Levine J, *et al.* Residual insulin production and pancreatic ß-cell turnover after 50 years of diabetes: Joslin Medalist Study. Diabetes 2010; 59(11): 2846-53.
[http://dx.doi.org/10.2337/db10-0676] [PMID: 20699420]

[4] Meier JJ, Bhushan A, Butler AE, Rizza RA, Butler PC. Sustained beta cell apoptosis in patients with long-standing type 1 diabetes: indirect evidence for islet regeneration? Diabetologia 2005; 48(11): 2221-8.
[http://dx.doi.org/10.1007/s00125-005-1949-2] [PMID: 16205882]

[5] Willcox A, Richardson SJ, Bone AJ, Foulis AK, Morgan NG. Evidence of increased islet cell proliferation in patients with recent-onset type 1 diabetes. Diabetologia 2010; 53(9): 2020-8.
[http://dx.doi.org/10.1007/s00125-010-1817-6] [PMID: 20532863]

[6] Menge BA, Tannapfel A, Belyaev O, *et al.* Partial pancreatectomy in adult humans does not provoke β-cell regeneration. Diabetes 2008; 57(1): 142-9.
[http://dx.doi.org/10.2337/db07-1294] [PMID: 17959931]

[7] Menge BA, Breuer TG, Ritter PR, Uhl W, Schmidt WE, Meier JJ. Long-term recovery of β-cell function after partial pancreatectomy in humans. Metabolism 2012; 61(5): 620-4.
[http://dx.doi.org/10.1016/j.metabol.2011.09.019] [PMID: 22079939]

List of Contributors

Aaron R. Cox McNair Medical Institute, Baylor College of Medicine, Houston, Texas

Amarnadh Nalla Department of Biomedical Sciences, University of Copenhagen, Denmark
Centre for Fetal Programming, Copenhagen, Denmark

Ananta Poudel Department of Medicine, The University of Chicago, Chicago, USA

Angela H. Darmon INSERM, U1016, Institut Cochin, Paris, France
Université Paris Descartes, Sorbonne Paris Cité, Faculté de Médecine, Paris, France

Benjamin Broche INSERM, U1016, Institut Cochin, Paris, France
Université Paris Descartes, Sorbonne Paris Cité, Faculté de Médecine, Paris, France

Bertrand Duvillié INSERM, U1016, Institut Cochin, Paris, France
Université Paris Descartes, Sorbonne Paris Cité, Faculté de Médecine, Paris, France

Birgitte Søstrup Department of Biomedical Sciences, University of Copenhagen, Denmark
Centre for Fetal Programming, Copenhagen, Denmark

Brian T. Layden Division of Endocrinology, Metabolism and Molecular Medicine, Northwestern University Feinberg School of Medicine, Chicago, Illinois, USA
Jesse Brown Veterans Affairs Medical Center, Chicago, Illinois, USA

Christine A. Beamish Department of Surgery, Islet Transplantation Laboratory, The Methodist Hospital Research Institute, Houston Texas, USA

David A. Hess Molecular Medicine Research Group, Krembil Centre for Stem Cell Biology, Robarts Research Institute; Department of Physiology & Pharmacology, Schulich School of Medicine & Dentistry, Western University, London, Ontario, Canada

David J. Hill Lawson Health Research Institute, St. Joseph's Health Care, 268 Grosvenor Street, London, Ontario N6A 4V2, Canada

Jeanette Kirkegaard Department of Biomedical Sciences, University of Copenhagen, Denmark
Novo Nordisk A/S, Måløv, Denmark

Jens H. Nielsen Department of Biomedical Sciences, University of Copenhagen, Denmark
Centre for Fetal Programming, Copenhagen, Denmark

Jonas Fowler Department of Medicine, The University of Chicago, Chicago, USA

Manami Hara Department of Medicine, The University of Chicago, Chicago, USA

Mulchand S. Patel Department of Biochemistry, School of Medicine and Biomedical Sciences, University at Buffalo, The State University of New York, Buffalo, New York 14214, USA

Paul Richards	INSERM, U1016, Institut Cochin, Paris, France Université Paris Descartes, Sorbonne Paris Cité, Faculté de Médecine, Paris, France
Ruth M. Elgamal	Molecular Medicine Research Group, Krembil Centre for Stem Cell Biology, Robarts Research Institute; Department of Physiology & Pharmacology, Schulich School of Medicine & Dentistry, Western University, London, Ontario, Canada
Saleh Mahmood	Department of Biochemistry, School of Medicine and Biomedical Sciences, University at Buffalo, The State University of New York, Buffalo, New York 14214, USA
Signe Horn	Department of Biomedical Sciences, University of Copenhagen, Denmark Novo Nordisk A/S, Måløv, Denmark
Stephanie Villa	Division of Endocrinology, Metabolism and Molecular Medicine, Northwestern University Feinberg School of Medicine, Chicago, Illinois, USA
Tyler T. Cooper	Molecular Medicine Research Group, Krembil Centre for Stem Cell Biology, Robarts Research Institute; Department of Physiology & Pharmacology, Schulich School of Medicine & Dentistry, Western University, London, Ontario, Canada
William L. Lowe	Division of Endocrinology, Metabolism and Molecular Medicine, Northwestern University Feinberg School of Medicine, Chicago, Illinois, USA

Current and Future Developments in Physiology
(Volume 1)
Control of Pancreatic Beta Cell Function and Plasticity in Health and Diabetes

Current and Future Developments in Physiology

Volume # 1

Control of Pancreatic Beta Cell Function and Plasticity in Health and Diabetes

Editor: David J. Hill

ISSN (print): 2468-7537

ISSN (online): 2468-7545

ISBN (online): 978-1-68108-365-0

ISBN (print): 978-1-68108-366-7

©2016, Bentham eBooks imprint.

Published by Bentham Science Publishers – Sharjah, UAE. All Rights Reserved.

CHAPTER 1

Understanding the Developmental Biology of β-Cells as a Strategy for Diabetes Reversal

Bertrand Duvillié[1,2,*], **Benjamin Broche**[1,2], **Angela Herengt Darmon**[1,2] and **Paul Richards**[1,2]

[1] *INSERM, U1016, Institut Cochin, Paris, France*

[2] *Université Paris Descartes, Sorbonne Paris Cité, Faculté de Médecine, Paris, France*

Abstract: In recent decades, intense efforts have been made to understand the cellular and molecular mechanisms controlling β-cell development. This process is well coordinated and consists of multiple steps. Many studies have tried to identify (i) molecular signals governing the proliferation of progenitors and (ii) their differentiation into mature pancreatic β-cells. A number of laboratories have focused on the role of transcription factors, and well constructed experiments have contributed to defining a hierarchy, highlighting the importance of each transcription factor in the interconnected network. Moreover, studies over the last 10 years have shown that the pancreatic mesenchymal cells, which are in contact with progenitors, influence pancreas organogenesis. Recent work has also indicated that the intra-uterine milieu influences gene expression and endocrine development. Indeed, nutrients, locally expressed growth factors and even the partial pressure of oxygen also control pancreas development. In a more applied setting, these understandings may improve our knowledge on the different forms of diabetes and, importantly, allow us to mimic a similar developmental process *in vitro*. This is because the precise understanding of each step *in vivo* seems to be necessary for designing protocols to generate β-cells from embryonic stem (ES) cells or induced pluripotent stem cells (iPS). These stem cell-derived β-cells should, in theory, provide new sources of insulin-secreting cells for transplantation into diabetic patients. A description of the recent advances in the field will be presented and illustrated in this chapter.

* **Corresponding author Bertrand Duvillié:** Institut Curie, Centre de Recherche, Laboratoire Eychene, INSERM U1021, CNRS UMR3347, Bât. 110, Centre Universitaire, 91405 ORSAY Cedex, France; Tel: (33) 1 69 86 30 74, Fax: (33) 1 69 86 30 51; E-mail: bertrand.duvillie@curie.fr

David J. Hill (Ed.)

Keywords: Beta cell, Development, Differentiation, FGF, HIF, Insulin, Oxygen, Pancreas, Proliferation, Reactive oxygen species.

GENETIC NETWORK CONTROLLING PANCREATIC DEVELOPMENT

The pancreas originates from the dorsal and ventral regions of the foregut endoderm, located directly behind the stomach. Signals from adjacent mesodermal structures, notochord, dorsal aorta and cardiac mesoderm are important for the emergence and early development of the pancreas [1 - 5]. During these early stages, *i.e.* E11.5 in the mouse, the pancreas is composed of an undifferentiated epithelium surrounded by mesenchyme. The epithelial fraction contains all the precursor cells that develop into the exocrine and endocrine compartments. A sequential expression of transcription factors is required to determine the commitment of these early progenitors to the different endocrine and exocrine cell types.

The Sequential Implication of Transcription Factors

These precursor cells express the gene *Pancreatic and Duodenal Factor 1* (*Pdx1*). This is a master gene as its deletion leads to the complete agenesis of the pancreas in the mouse. In humans, the paralogue of Pdx1 is called insulin promoting factor 1 (IPF1). Interestingly, a pancreatic agenesis was also described in a patient with a homozygous single nucleotide deletion in the codon 63 of IPF1. The role of Pdx1 is thus conserved between rodents and humans. Following Pdx1 expression, the transcription factor Neurogenin 3 (NGN3) is transiently expressed in endocrine progenitors. The disruption of *Ngn3* results in the absence of α–, β–, δ– and PP-cells [6]. Other transcription factors are involved more specifically in exocrine cell development, for example MIST1 [7]. The genetic network controlling pancreas development is represented in Fig. (**1**).

In addition to these complex genetic mechanisms, it has been shown that the local environment releases signals that determine not only proliferation of the precursor cells but also their differentiation. A detailed review of these signals will be presented in the next section.

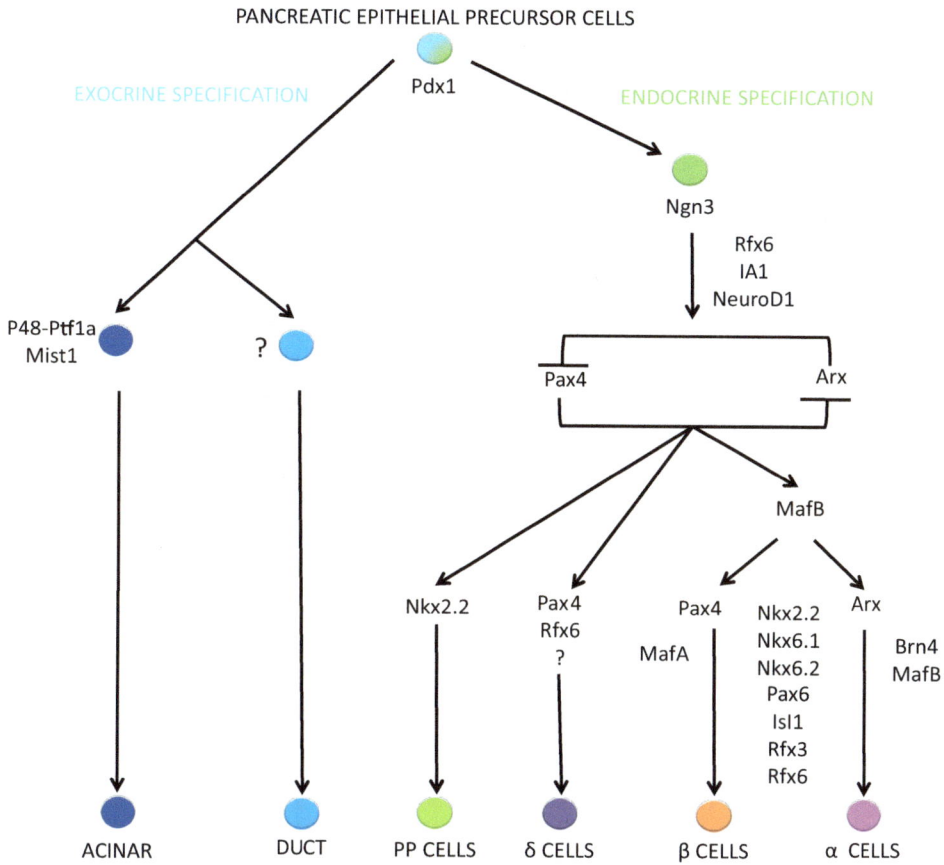

Fig. (1). The genetics of pancreas development [8].

THE PANCREATIC MESENCHYME AND GROWTH FACTORS

For many years the exact role of the mesenchymal cells that surround the epithelial precursors remained elusive and was the subject of much debate. In the laboratory of R Scharfmann, an *in vitro* model allowing the development of the embryonic pancreas in the presence or absence of the mesenchyme was designed [9]. In these conditions, more acinar cells developed in the presence rather than in the absence of the mesenchyme. To investigate the mechanism by which the mesenchyme exerts its effects, the expression of tyrosine kinase receptors was researched. A special interest was paid to the Fibroblast Growth Factors family (FGFs). There are four FGF receptors, called FGF1 to FGF4, with several isoforms. Of importance, the FGFR2IIIb isoform expression is specific to the epithelium, and can bind several FGFs, including FGF 1, FGF 3, FGF 7, FGF 10 and FGF 22 [10, 11]. During pancreas development, the receptor FGFR2IIIb was found to be expressed in the epithelial precursor cells. Moreover, its ligands FGF1, FGF7 and FGF10 were also expressed in the pancreas during this period. Interestingly, *in situ* hybridization showed that FGF10 is expressed in the mesenchyme directly adjacent to the ventral and dorsal pancreatic buds. To investigate the role of FGF10, knock-out animals were produced and analyzed. At the end of gestation, the size of the pancreas of the FGF10-/- embryos was considerably reduced [12], highlighting the importance of FGF10 for the development of pancreas. In these animals, at early stages of development (E9.5-E10.5), the emergence of the epithelium within the dorsal and ventral pancreatic buds was normal. However, the growth, differentiation and branching of the epithelial were arrested. These defects were caused by a dramatic reduction of the proliferation of the precursor cells expressing PDX1 [12]. Thus, these genetic experiments in mice demonstrated that FGF10 expression is mandatory for the harmonious development of the pancreas. Two other laboratories showed that misexpression of FGF10 in the PDX1-positive cells blocked β-cell development by stimulating the Notch pathway [13, 14]. Such results seemed to be contradictory with the knock-out experiments. To clarify the exact role of FGF10 during development, we performed other gain of function experiments using another *in vitro* model, in which the pancreas is cultured on a filter at the air/liquid interface. We first compared the *in vitro* and *in vivo* sequence of gene expression and found that this model replicates all the steps of pancreatic development from the progenitors [15]. When the embryonic pancreatic epithelium was cultured in the presence of the mesenchyme or FGF10, the proliferation of the early PDX1-positive progenitors was increased compared to the epithelium alone. Consequently, the number of endocrine progenitors expressing NGN3 was elevated. Finally, the number of differentiated β-cells was also enhanced by FGF10 or the mesenchyme [15]. This mechanism mediated by FGF10 had also been observed using human embryonic pancreatic cells [16]. FGF7, another ligand of FGFR2IIIB receptor, also has the capability to stimulate the proliferation of the precursor cells. Thus, these studies demonstrate that mesenchymal signals, and particularly FGF10, can be used to stimulate progenitor cell multiplication and their later differentiation into β-cells. Recently, growth factors have been used to expand human progenitor cells in a 3D culture system. The groups of Scharfmann and Heimberg used a combination of the Wnt agonist R-Spondin 1, FGF10 and epidermal growth factor (EGF). This cocktail allowed the amplification of the progenitor cells. Interestingly, in the absence of EGF, proliferation decreased while endocrine differentiation was upregulated [17]. Thus, these experiments highlight the importance of EGF to manipulate the balance between proliferation and differentiation of progenitors.

ROLE OF NUTRIENTS IN THE DEVELOPMENT OF B-CELLS

In order to evaluate the impact of nutrition on β-cell development, nutrient restriction has been applied to pregnant rats from the last week of gestation until the weaning of the pups. The β-cell mass of the pups was measured using quantitative morphometry analysis. In this experimental model, the fetuses were irreversibly growth retarded and their pancreases contained reduced β-cell mass compared to the controls. This defect in β-cell mass persisted into adulthood and led to glucose intolerance [18, 19]. Moreover, in adulthood, the females had an impaired adaptation of β-cell mass to gestation [20] and their fetuses (second generation) also displayed some defects in β-cell development [21]. By comparison, instead of restricting food to rats, other studies have used a hypocaloric regimen in mice during the last week of gestation. In these studies, the weight of the newborns was reduced at birth but was normalized at later stages. Moreover, the β-cell mass was normal during adulthood, but the animals became glucose-intolerant due to defects in insulin-secretion [22].

Impact of Protein Levels

Protein isocaloric restriction during pregnancy and lactation causes a mild intrauterine growth retardation and alterations in pancreas development [23, 24]. Interestingly, a reduction of dietary protein during gestation results in reduced pancreas weight of neonates. These neonates have reduced β-cell mass, islet vascularization, diminished β-cell replication and increased apoptosis [25, 26]. More recently, the role of Leucine, a branched essential amino acid, was analyzed. Leucine is known to activate the mammalian target of rapamycine (mTOR) signaling pathway. Rachdi *et al.* showed that Leucine supplementation during pregnancy increased fetal body weight, and caused fetal hyperglycemia, hypoinsulinemia and decreased islet area [27]. In culture, addition of Leucine altered the differentiation of PDX1+ precursor cells into Neurogenin 3-expressing cells. At the molecular level, it was found that Leucine supplementation favored the Hypoxia Inducible Factor 1-alpha stabilization, which is a repressor of β-cell development. These findings suggest that Leucine supplementation during pregnancy may increase the susceptibility of individuals to Type 2 diabetes.

The Effects of Glucose

To investigate the role of glucose, the *in vitro* culture system described previously, in which the pancreatic buds are cultured at the air/liquid interface, was used [15]. Rat embryonic pancreases were cultured in the presence or in the absence of glucose [28]. The proliferation of progenitors, cell survival and acinar cell development were not found to be dependent on the presence or absence of glucose. β-cell development was, however, regulated by glucose in a dose-dependent manner. In this study, glucose was shown to control the expression of the transcription factor Neuro D, which is a target of NGN3. By regulating this genetic cascade, glucose seems to be an important regulator of endocrine differentiation.

IMPORTANCE OF THE PARTIAL PRESSURE OF OXYGEN IN THE INTRAUTERINE MILIEU

During the last few years, oxygen tension has been shown to play an important role during pancreas development. Indeed, during early fetal development there is a paucity of blood flow, causing hypoxia. Later, the blood flow and partial pressure of oxygen increases in the pancreas [29] in parallel to β-cell development. Using different experimental models, we have shown that oxygen tension tightly controls β-cell differentiation [8, 30]. Indeed, when undifferentiated rat pancreases are cultured in hypoxic conditions (pO_2=3%), only rare β-cells develop. On the other hand, increasing pO_2 to 21% enhance β-cell differentiation by activating Ngn3 expression. Such results are consistent with the findings from other laboratories [31]. Most of the important responses in cells to hypoxia are mediated by the transcription factor "Hypoxia Inducible Factor 1 alpha" (HIF1-alpha). The stabilization of HIF1-alpha during hypoxia is dependent on the activity of Prolyl Hydroxylases (PHDs). In the presence of oxygen PHDs are active and they hydroxylate two proline residues of HIF1-alpha. Such modifications allow its recognition by the von Hippel Lindau E3 ubiquitin ligase complex, leading to its degradation by the proteasome. In hypoxia, the activity of PHDs is reduced and HIF is stabilized. When HIF1-alpha was stabilized chemically using PHD inhibitors, β-cell differentiation was reduced *in vitro* [30]. Moreover, genetic deletion of Vhl in the progenitor cells *in vivo* leads to constitutive HIF1-alpha stabilization, a decrease of Ngn3 expression, and a dramatic reduction of β-cell development [32]. Together, these findings strongly support the notion that HIF has a negative action on β-cell differentiation (Fig. **2**). One thing that still needs to be elucidated, however, is by which mechanism HIF exerts such negative effects on endocrine differentiation. We have analyzed the hypothesis of a vascular effect, as HIF1-alpha is an activator of the Vascular Endothelial Growth Factor expression. Moreover, it was recently shown that hyper-vascularization during the fetal life can restrict pancreas development [33]. We first found that vhl mutant fetal pancreases had increased VEGF expression and the number of blood vessels was elevated when compared to controls. Next, we used an *in vitro* model in which the vhl ko mutants developed similarly to *in vivo*, *i.e.* reduced β-cell differentiation and increased vascularization. By using an inhibitor of the VEGF receptor (VEGFR2), we could then reduce the number of blood vessels and consequently the development of β-cells was restored. Together, these data suggest that during the fetal pancreas development, the restrictive effect of HIF1-alpha on β-cells development is mediated by blood vessels.

OXIDATIVE STRESS AND B-CELL DEVELOPMENT

In general, oxidative signals seem important for stem cells biology. Oxidative stress is an imbalance between the production of reactive oxygen species (ROS) and their detoxification by the cell. It results in an increased concentration of ROS that can lead to toxic effects. On the other hand, several laboratories have shown that lower levels of ROS can also control the proliferation, survival, and differentiation of the cells [34 - 36]. These mechanisms have been illustrated by using several different cell types. For example, the hematopoietic stem cells are quiescent when the levels of ROS are low. On the contrary, they start proliferating rapidly and finally they exit their cycles of their self-renewal when the levels of ROS increase [37]. These observations suggest that both stemness and quiescence are associated with low levels of ROS in stem cell niches.

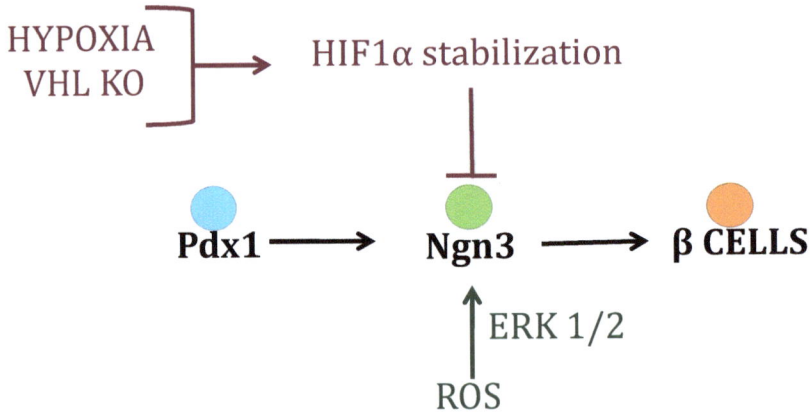

Fig. (2). Control of β-cell development by oxygen and ROS levels.

Recently, we examined the expression of antioxidant enzymes genes in the embryonic and adult pancreas, compared to the liver. In particular we focused on catalase and glutathione peroxidase,. These two enzymes control the degradation of H_2O_2. Expression levels were very weak in the adult pancreas as compared to the liver, and were even lower in the embryonic pancreas [38]. This suggests that the pancreatic progenitors may be highly susceptible to H_2O_2 accumulation, which can in turn activate signaling pathways involved in cell proliferation and differentiation [39]. To research the effects of ROS in pancreatic development, we first treated pregnant rats with the antioxydant N-Acetyl Cystein (NAC). Such treatment decreased β-cell development, indicating that ROS are probably involved in this process. Next, we used culture models. When undifferentiated E13.5 pancreases were treated with H_2O_2, the differentiation of β-cells was increased, through the augmentation of NGN3 expression (Fig. **2**). On the other hand, when ROS levels were reduced by overexpressing catalase, or by using NAC or m-chlorophenyl hydrazone (CCCP) treatments, β-cell differentiation was reduced. Finally, we showed that the ERK1/2 was activated by H_2O_2, and the differentiation of β-cells was blocked when the embryonic pancreases were exposed to an ERK1/2 inhibitor. Together, these data strongly support the fact that

ROS are important signals for endocrine development.

The positive roles of ROS in the pancreas have also been further illustrated in the adult mature β-cells. Leloup *et al.* showed that decreasing ROS in rat islets by using trolox, another antioxidant, reduces glucose stimulated insulin secretion [40]. Thus, despite the fact that in a limited energy environment, mitochondrial dysfunction can lead to overproduction of reactive oxygen species and to a decline of β-cell function, we think that it is also important to consider the ROS for their positive effects as messengers.

FROM DEVELOPMENTAL BIOLOGY STUDIES TO INNOVATIVE THERAPEUTIC STRATEGIES FOR THE TREATMENT OF DIABETES

The Context of Diabetes

Diabetes affects more than 300 millions individuals. It is also responsible for 1.5 million deaths per year. Today, patients affected by different types of diabetes can be treated using insulin injections or antidiabetic drugs, but there is no long-term cure. Better treatments for diabetes thus represent a major therapeutic and economic challenge.

Diabetes leads to hyperglycemia and abnormal β-cell function. Type-2 diabetes is the most frequent form and it is a multifactorial disease, modulated by several susceptibility genes and different factors, such as excessive consumption of saturated fatty acids, quick-burning sugars and having a sedate lifestyle. In addition to the genetic predisposition to diabetes, an increasing amount of data has shed light on the role of the fetal environment on the post-natal life of β-cells [41, 42]. Indeed, fetal environmental stress leads to the appearance of severe diseases such as diabetes and obesity [43].

Application of Our Knowledge From β-Cell Development to Treating Diabetes

Type 1 Diabetes and Islet Transplantation

Type 1 diabetes (T1D), also named insulino-dependent diabetes, is an autoimmune disease that represents 10% of all diabetic patients. The cause of the disease is not fully understood but it leads to β-cell destruction and, consequently, an inadequate control of blood glucose. In general, these patients are treated with replacement insulin during meals, and to date, it remains a chronic disease.

New strategies have been developed to preserve β-cells. The advantage of this is that it can help avoid several side effects of diabetes, including detrimental alterations to the heart, kidney disease and retinopathy. Rather than using chronic treatments though, the graft of human β-cells have been investigated. Shapiro *et al.* [44], in their original study, obtained human islets from post-mortem donors and introduced them in to the portal vein. With this protocol, the islets circulate to and then reside in the liver. A glucocorticoid-free immunosuppresive regimen consisting of sirolimus, tacrolimus and daclizumab was administrated to the grafted patients. Following this, most of the patients became normoglycemic and independent of insulin injections. A five-year follow up of this clinical study, however, indicated that the treatment gradually loss its efficiency [45]. Since this first trial, considerable efforts have been performed worldwide to increase the efficiency of such procedures, notably in terms of the duration of graft survival and function. Vantyghem *et al.* analyzed the importance of the primary graft function on graft survival and activity [46]. In particular, after two or three sequential islets infusions, the primary graft function was estimated by measuring the beta score, a validated index based on insulin treatment requirement, plasma C peptide, blood glucose and HbA1c. After a 3.3 years follow-up, 57% (n=8) of the patients remained insulin independent with HbA1C < 6%, including 78% of them who had optimal primary graft function [46]. These observations indicate that optimal primary graft function prolongs the graft survival and function. The major challenge facing the widespread use of islet transplantation, however, is the poor availability of islets from human donors [47]. This has prompted the research into other sources of pancreatic islets. Among non-human islets, the most relevant candidate so far are porcine islets. However, the possibility of xenozoonosis has delayed clinical trials. A long term follow-up study will eventually need to be performed to monitor the immune rejections to these grafts too. Alternative strategies have also been used to produce human β-cells from other sources, including embryonic stem cells, adult stem cells, or differentiated cells of other types that can be directed toward an insulin-producing cell fate.

Whatever the selected strategy, understanding the cellular and molecular mechanisms that govern β-cell development, and their preservation during embryogenesis, appear to be key challenges. In particular, we think that the increased knowledge of these mechanisms will allow possibilities to better generate β-cells *in vitro* and *in vivo*. Indeed, a number of discoveries obtained from developmental biology studies have been applied to improve protocols for generating β-cells from ES cells. This is the case, for example, with FGF10. As described previously, FGF10 increases the proliferation of pancreatic progenitors in rodents and humans. On the basis of such results, FGF10 was used to derive β-like cells from ES cells [48]. These protocols allowed the efficient generation of pancreatic endoderm and pancreatic progenitors that can be differentiated into functional β-cells within 3-4 weeks following their transplantation into rodents [49, 50]. More recently, the laboratory of Melton reported the discovery of a large-scale production of functional β-cells from human pluripotent stem cells *in vitro* [51]. Using a sequential modulation of multiple signaling pathways in 3D cultures, they obtained glucose-responsive mono-hormonal insulin producing cells, which resemble those found in human islets. In such protocols, another member of the FGF family, FGF7 (also called Keratinocyte Growth Factor), is used. This factor has also been described previously in developmental biology studies [16]. Finally, it was shown that high oxygen conditions facilitate the differentiation of mouse and human pluripotent stem cells into pancreatic progenitors and insulin producing cells [52], on the basis that oxygen tension tightly controls β-cells differentiation [30, 53]. Altogether, these new improvements of cell therapy are providing efficient ways to increase the derivation of β-cells *in vitro* and potentially use them to treat diabetes.

THE GENETICS OF PANCREAS EMBRYOGENESIS AND THE OCCURRENCE OF DIABETES

Developmental genes have been implicated in a number of different forms of diabetes. This is the case, for example, with several types of Maturity Onset Diabetes of the Young (MODY). MODY is described as several hereditary forms of diabetes caused by mutations in an autosomal dominant gene, leading to insulin disruption. MODY3, MODY4 and MODY9 are caused by mutations of Heaptocyte Nuclear 1 alpha (HNF1α), IPF1 (called PDX1 in rodents) and PAX4 respectively [54]. Moreover, recent investigations were performed to identify the role of "developmental" genes in the function of mature β-cells. One recent example is that of studies investigating the transcription factor Rfx6. Rfx6 was first shown to direct islet cell differentiation downstream of Ngn3 [55] and mice lacking Rfx6 did not produce pancreatic endocrine hormones, except for pancreatic-polypeptide. Human infants with mutations in the Rfx6 gene have a severe form neonatal diabetes that requires constant insulin treatment due to the lack of the hormone in the circulation. So, taken together Rfx6 can be considered a critical transcription factor for endocrine pancreas development. Its function in mature β-cells has, however, remained elusive until recently. To shed light on this, a study used a human β-cell line and islets from adult human donors. They found Rfx6 controls insulin expression, content and Ca(2+) flux dependent insulin secretion in mature cells [56]. β-cells containing a missense mutation of RFX6, associated with neonatal diabetes, lacked the capacity to transduce insulin expression and to increase Ca(2+) channel expression [56]. Together, these results highlight functional roles of RFX6 at the cellular level.

In conclusion, the study of pancreatic development has several applications. First, the identification of genetic and environmental factors that control β-cell differentiation can be applied to derive new β-cells *in vitro* from ES or iPS cells. Second, it may help to better understand the physiopathology of β-cells. Third, accumulation of data on pancreas development may allow a better understanding of how to induce β-cell regeneration in diabetic adults.

CONFLICT OF INTEREST

The authors confirm that they have no conflict of interest to declare for this publication.

ACKNOWLEDGEMENTS

Declared none.

REFERENCES

[1]   Deutsch G, Jung J, Zheng M, Lóra J, Zaret KS. A bipotential precursor population for pancreas and liver within the embryonic endoderm. Development 2001; 128(6): 871-81.
[PMID: 11222142]

[2]   Lammert E, Cleaver O, Melton D. Induction of pancreatic differentiation by signals from blood vessels. Science 2001; 294(5542): 564-7.
[http://dx.doi.org/10.1126/science.1064344] [PMID: 11577200]

[3]   Kim SK, Hebrok M, Melton DA. Pancreas development in the chick embryo. Cold Spring Harb Symp Quant Biol 1997; 62: 377-83.
[http://dx.doi.org/10.1101/SQB.1997.062.01.045] [PMID: 9598372]

[4]   Kim SK, Hebrok M. Intercellular signals regulating pancreas development and function. Genes Dev 2001; 15(2): 111-27.
[http://dx.doi.org/10.1101/gad.859401] [PMID: 11157769]

[5]   Hebrok M, Kim SK, Melton DA. Notochord repression of endodermal Sonic hedgehog permits pancreas development. Genes Dev 1998; 12(11): 1705-13.
[http://dx.doi.org/10.1101/gad.12.11.1705] [PMID: 9620856]

[6]   Gradwohl G, Dierich A, LeMeur M, Guillemot F. neurogenin3 is required for the development of the four endocrine cell lineages of the pancreas. Proc Natl Acad Sci USA 2000; 97(4): 1607-11.
[http://dx.doi.org/10.1073/pnas.97.4.1607] [PMID: 10677506]

[7]   Pin CL, Rukstalis JM, Johnson C, Konieczny SF. The bHLH transcription factor Mist1 is required to maintain exocrine pancreas cell organization and acinar cell identity. J Cell Biol 2001; 155(4): 519-30.
[http://dx.doi.org/10.1083/jcb.200105060] [PMID: 11696558]

[8]   Heinis M, Simon MT, Duvillié B. New insights into endocrine pancreatic development: the role of environmental factors. Horm Res Paediatr 2010; 74(2): 77-82.
[http://dx.doi.org/10.1159/000314894] [PMID: 20551619]

[9]   Miralles F, Czernichow P, Scharfmann R. Follistatin regulates the relative proportions of endocrine *versus* exocrine tissue during pancreatic development. Development 1998; 125(6): 1017-24.
[PMID: 9463348]

[10] Ornitz DM, Xu J, Colvin JS, *et al.* Receptor specificity of the fibroblast growth factor family. J Biol Chem 1996; 271(25): 15292-7.
[http://dx.doi.org/10.1074/jbc.271.25.15292] [PMID: 8663044]

[11] Zhang X, Ibrahimi OA, Olsen SK, Umemori H, Mohammadi M, Ornitz DM. Receptor specificity of the fibroblast growth factor family. The complete mammalian FGF family. J Biol Chem 2006; 281(23): 15694-700.
[http://dx.doi.org/10.1074/jbc.M601252200] [PMID: 16597617]

[12] Bhushan A, Itoh N, Kato S, *et al.* Fgf10 is essential for maintaining the proliferative capacity of epithelial progenitor cells during early pancreatic organogenesis. Development 2001; 128(24): 5109-17.
[PMID: 11748146]

[13]   Hart A, Papadopoulou S, Edlund H. Fgf10 maintains notch activation, stimulates proliferation, and blocks differentiation of pancreatic epithelial cells. Dev Dyn 2003; 228(2): 185-93.
[http://dx.doi.org/10.1002/dvdy.10368] [PMID: 14517990]

[14]   Norgaard GA, Jensen JN, Jensen J. FGF10 signaling maintains the pancreatic progenitor cell state revealing a novel role of Notch in organ development. Dev Biol 2003; 264(2): 323-38.
[http://dx.doi.org/10.1016/j.ydbio.2003.08.013] [PMID: 14651921]

[15]   Attali M, Stetsyuk V, Basmaciogullari A, *et al.* Control of beta-cell differentiation by the pancreatic mesenchyme. Diabetes 2007; 56(5): 1248-58.
[http://dx.doi.org/10.2337/db06-1307] [PMID: 17322477]

[16]   Ye F, Duvillié B, Scharfmann R. Fibroblast growth factors 7 and 10 are expressed in the human embryonic pancreatic mesenchyme and promote the proliferation of embryonic pancreatic epithelial cells. Diabetologia 2005; 48(2): 277-81.
[http://dx.doi.org/10.1007/s00125-004-1638-6] [PMID: 15690149]

[17]   Bonfanti P, Nobecourt E, Oshima M, *et al. Ex Vivo* Expansion and differentiation of human and mouse fetal pancreatic progenitors are modulated by epidermal growth factor. Stem Cells Dev 2015; 24(15): 1766-78.
[http://dx.doi.org/10.1089/scd.2014.0550] [PMID: 25925840]

[18]   Garofano A, Czernichow P, Bréant B. Effect of ageing on beta-cell mass and function in rats malnourished during the perinatal period. Diabetologia 1999; 42(6): 711-8.
[http://dx.doi.org/10.1007/s001250051219] [PMID: 10382591]

[19]   Garofano A, Czernichow P, Bréant B. In utero undernutrition impairs rat beta-cell development. Diabetologia 1997; 40(10): 1231-4.
[http://dx.doi.org/10.1007/s001250050812] [PMID: 9349607]

[20]   Blondeau B, Garofano A, Czernichow P, Bréant B. Age-dependent inability of the endocrine pancreas to adapt to pregnancy: a long-term consequence of perinatal malnutrition in the rat. Endocrinology 1999; 140(9): 4208-13.
[PMID: 10465293]

[21]   Blondeau B, Avril I, Duchene B, Bréant B. Endocrine pancreas development is altered in foetuses from rats previously showing intra-uterine growth retardation in response to malnutrition. Diabetologia 2002; 45(3): 394-401.
[http://dx.doi.org/10.1007/s00125-001-0767-4] [PMID: 11914745]

[22]   Jimenez-Chillaron JC, Hernandez-Valencia M, Reamer C, *et al.* Beta-cell secretory dysfunction in the pathogenesis of low birth weight-associated diabetes: a murine model. Diabetes 2005; 54(3): 702-11.
[http://dx.doi.org/10.2337/diabetes.54.3.702] [PMID: 15734846]

[23]   Reusens B, Remacle C. Programming of the endocrine pancreas by the early nutritional environment. Int J Biochem Cell Biol 2006; 38(5-6): 913-22.
[http://dx.doi.org/10.1016/j.biocel.2005.10.012] [PMID: 16337425]

[24]   Hales CN, Ozanne SE. For debate: Fetal and early postnatal growth restriction lead to diabetes, the metabolic syndrome and renal failure. Diabetologia 2003; 46(7): 1013-9.
[http://dx.doi.org/10.1007/s00125-003-1131-7] [PMID: 12827239]

[25] Petrik J, Reusens B, Arany E, *et al.* A low protein diet alters the balance of islet cell replication and apoptosis in the fetal and neonatal rat and is associated with a reduced pancreatic expression of insulin-like growth factor-II. Endocrinology 1999; 140(10): 4861-73.
[PMID: 10499546]

[26] Boujendar S, Arany E, Hill D, Remacle C, Reusens B. Taurine supplementation of a low protein diet fed to rat dams normalizes the vascularization of the fetal endocrine pancreas. J Nutr 2003; 133(9): 2820-5.
[PMID: 12949371]

[27] Rachdi L, Aïello V, Duvillié B, Scharfmann R. L-leucine alters pancreatic β-cell differentiation and function *via* the mTor signaling pathway. Diabetes 2012; 61(2): 409-17.
[http://dx.doi.org/10.2337/db11-0765] [PMID: 22210321]

[28] Guillemain G, Filhoulaud G, Da Silva-Xavier G, Rutter GA, Scharfmann R. Glucose is necessary for embryonic pancreatic endocrine cell differentiation. J Biol Chem 2007; 282(20): 15228-37.
[http://dx.doi.org/10.1074/jbc.M610986200] [PMID: 17376780]

[29] Shah SR, Esni F, Jakub A, *et al.* Embryonic mouse blood flow and oxygen correlate with early pancreatic differentiation. Dev Biol 2011; 349(2): 342-9.
[http://dx.doi.org/10.1016/j.ydbio.2010.10.033] [PMID: 21050843]

[30] Heinis M, Simon MT, Ilc K, *et al.* Oxygen tension regulates pancreatic beta-cell differentiation through hypoxia-inducible factor 1alpha. Diabetes 2010; 59(3): 662-9.
[http://dx.doi.org/10.2337/db09-0891] [PMID: 20009089]

[31] Fraker CA, Ricordi C, Inverardi L, Domínguez-Bendala J. Oxygen: a master regulator of pancreatic development? Biol Cell 2009; 101(8): 431-40.
[http://dx.doi.org/10.1042/BC20080178] [PMID: 19583566]

[32] Soggia A, Ramond C, Akiyama H, Scharfmann R, Duvillie B. von Hippel-Lindau gene disruption in mouse pancreatic progenitors and its consequences on endocrine differentiation *in vivo*: importance of HIF1-α and VEGF-A upregulation. Diabetologia 2014; 57(11): 2348-56.
[http://dx.doi.org/10.1007/s00125-014-3365-y] [PMID: 25186293]

[33] Magenheim J, Ilovich O, Lazarus A, *et al.* Blood vessels restrain pancreas branching, differentiation and growth. Development 2011; 138(21): 4743-52.
[http://dx.doi.org/10.1242/dev.066548] [PMID: 21965615]

[34] Blanchetot C, Boonstra J. The ROS-NOX connection in cancer and angiogenesis. Crit Rev Eukaryot Gene Expr 2008; 18(1): 35-45. [Review].
[http://dx.doi.org/10.1615/CritRevEukarGeneExpr.v18.i1.30] [PMID: 18197784]

[35] Chiarugi P, Fiaschi T. Redox signalling in anchorage-dependent cell growth. Cell Signal 2007; 19(4): 672-82.
[http://dx.doi.org/10.1016/j.cellsig.2006.11.009] [PMID: 17204396]

[36] Leslie NR. The redox regulation of PI 3-kinase-dependent signaling. Antioxid Redox Signal 2006; 8(9-10): 1765-74.
[http://dx.doi.org/10.1089/ars.2006.8.1765] [PMID: 16987030]

[37] Jang YY, Sharkis SJ. A low level of reactive oxygen species selects for primitive hematopoietic stem

cells that may reside in the low-oxygenic niche. Blood 2007; 110(8): 3056-63.
[http://dx.doi.org/10.1182/blood-2007-05-087759] [PMID: 17595331]

[38]   Hoarau E, Chandra V, Rustin P, Scharfmann R, Duvillie B. Pro-oxidant/antioxidant balance controls pancreatic β-cell differentiation through the ERK1/2 pathway. Cell Death Dis 2014; 5: e1487.
[http://dx.doi.org/10.1038/cddis.2014.441] [PMID: 25341041]

[39]   Rhee SG, Kang SW, Jeong W, Chang TS, Yang KS, Woo HA. Intracellular messenger function of hydrogen peroxide and its regulation by peroxiredoxins. Curr Opin Cell Biol 2005; 17(2): 183-9.
[http://dx.doi.org/10.1016/j.ceb.2005.02.004] [PMID: 15780595]

[40]   Leloup C, Tourrel-Cuzin C, Magnan C, *et al.* Mitochondrial reactive oxygen species are obligatory signals for glucose-induced insulin secretion. Diabetes 2009; 58(3): 673-81.
[http://dx.doi.org/10.2337/db07-1056] [PMID: 19073765]

[41]   Barker DJ. Developmental origins of adult health and disease. J Epidemiol Communi Health 2004; 58(2): 114-5.
[http://dx.doi.org/10.1136/jech.58.2.114] [PMID: 14729887]

[42]   Barker DJ. The origins of the developmental origins theory. J Intern Med 2007; 261(5): 412-7.
[http://dx.doi.org/10.1111/j.1365-2796.2007.01809.x] [PMID: 17444880]

[43]   McMillen IC, Robinson JS. Developmental origins of the metabolic syndrome: prediction, plasticity, and programming. Physiol Rev 2005; 85(2): 571-633.
[http://dx.doi.org/10.1152/physrev.00053.2003] [PMID: 15788706]

[44]   Shapiro AM, Lakey JR, Ryan EA, *et al.* Islet transplantation in seven patients with type 1 diabetes mellitus using a glucocorticoid-free immunosuppressive regimen. N Engl J Med 2000; 343(4): 230-8.
[http://dx.doi.org/10.1056/NEJM200007273430401] [PMID: 10911004]

[45]   Ryan EA, Paty BW, Senior PA, *et al.* Five-year follow-up after clinical islet transplantation. Diabetes 2005; 54(7): 2060-9.
[http://dx.doi.org/10.2337/diabetes.54.7.2060] [PMID: 15983207]

[46]   Vantyghem MC, Kerr-Conte J, Arnalsteen L, *et al.* Primary graft function, metabolic control, and graft survival after islet transplantation. Diabetes Care 2009; 32(8): 1473-8.
[http://dx.doi.org/10.2337/dc08-1685] [PMID: 19638525]

[47]   Lechner A, Habener JF. Stem/progenitor cells derived from adult tissues: potential for the treatment of diabetes mellitus. Am J Physiol Endocrinol Metab 2003; 284(2): E259-66.
[http://dx.doi.org/10.1152/ajpendo.00393.2002] [PMID: 12531740]

[48]   DAmour KA, Bang AG, Eliazer S, *et al.* Production of pancreatic hormone-expressing endocrine cells from human embryonic stem cells. Nat Biotechnol 2006; 24(11): 1392-401.
[http://dx.doi.org/10.1038/nbt1259] [PMID: 17053790]

[49]   Kroon E, Martinson LA, Kadoya K, *et al.* Pancreatic endoderm derived from human embryonic stem cells generates glucose-responsive insulin-secreting cells *in vivo.* Nat Biotechnol 2008; 26(4): 443-52.
[http://dx.doi.org/10.1038/nbt1393] [PMID: 18288110]

[50]   Rezania A, Bruin JE, Riedel MJ, *et al.* Maturation of human embryonic stem cell-derived pancreatic progenitors into functional islets capable of treating pre-existing diabetes in mice. Diabetes 2012; 61(8): 2016-29.

[http://dx.doi.org/10.2337/db11-1711] [PMID: 22740171]

[51]   Pagliuca FW, Millman JR, Gürtler M, *et al*. Generation of functional human pancreatic β cells *in vitro*. Cell 2014; 159(2): 428-39.
[http://dx.doi.org/10.1016/j.cell.2014.09.040] [PMID: 25303535]

[52]   Hakim F, Kaitsuka T, Raeed JM, *et al*. High oxygen condition facilitates the differentiation of mouse and human pluripotent stem cells into pancreatic progenitors and insulin-producing cells. J Biol Chem 2014; 289(14): 9623-38.
[http://dx.doi.org/10.1074/jbc.M113.524363] [PMID: 24554704]

[53]   Heinis M, Soggia A, Bechetoille C, *et al*. HIF1α and pancreatic β-cell development. FASEB J 2012; 26(7): 2734-42.
[http://dx.doi.org/10.1096/fj.11-199224] [PMID: 22426121]

[54]   Steck AK, Winter WE. Review on monogenic diabetes. Curr Opin Endocrinol Diabetes Obes 2011; 18(4): 252-8.
[http://dx.doi.org/10.1097/MED.0b013e3283488275] [PMID: 21844708]

[55]   Smith SB, Qu HQ, Taleb N, *et al*. Rfx6 directs islet formation and insulin production in mice and humans. Nature 2010; 463(7282): 775-80.
[http://dx.doi.org/10.1038/nature08748] [PMID: 20148032]

[56]   Chandra V, Albagli-Curiel O, Hastoy B, *et al*. RFX6 regulates insulin secretion by modulating Ca2+ homeostasis in human β cells. Cell Reports 2014; 9(6): 2206-18.
[http://dx.doi.org/10.1016/j.celrep.2014.11.010] [PMID: 25497100]

CHAPTER 2

Aging and β-Cell Proliferation, Molecular and Signaling Changes and What This Means for Targeted Regeneration

Aaron R. Cox[*]

McNair Medical Institute, Baylor College of Medicine, Houston, Texas, U.S.A

Abstract: Increased age confers a greater risk for the development of type 2 diabetes (T2D), and also has significant consequences for β-cell growth and regeneration. Pancreatic insulin-producing β-cells are long-lived, and exhibit very little turnover in adult life. The severe decline in β-cell proliferation contributes to a decreased capacity for β-cell regeneration with age. β-cell regeneration is dependent on mitogenic signals, receptor and downstream signal transduction, cell cycle progression, and epigenetic regulation of gene expression, all of which are significantly affected by increasing age. Studies suggest that circulating growth factors and their receptors are decreased with age, along with important intracellular signaling molecules, such as Pdx-1 and FoxM1. Cell cycle progression is inhibited by an increased expression of cell cycle inhibitors and a reduction in cell cycle kinase complexes (Cyclin/Cdks). Moreover, decreased expression of epigenetic silencers, such as polycomb group proteins, results in de-repression of the cell cycle inhibitor p16, and a significant reduction in β-cell proliferation. Collectively, these age-induced changes present obstacles for the design of β-cell regenerative therapies for diabetes; however, some reports suggest that even very old β-cells can re-enter cell cycle. Future studies will further define the effects of aging on β-cell proliferation and elucidate new drug targets for diabetes therapy.

Keywords: Aging, β-cell regeneration, Cell cycle, Diabetes, Epigenetics, Molecular signals, Proliferation.

[*] **Corresponding author Aaron R. Cox:** McNair Medical Institute, Baylor College of Medicine, Houston, Texas, U.S.A; Tel/Fax: 832-824-0711; E-mail: racox@bcm.edu

David J. Hill (Ed.)

AGING AND DIABETES

Aging is associated with an increased risk for several diseases including metabolic syndrome and type 2 diabetes (T2D). The incidence of T2D increases with age from ~8% in middle age (40-59 years) to ~33% in older adults (>60 years) [1]. These rates increase substantially when accounting for individuals with prediabetes, defined as impaired fasting glucose or glucose tolerance. There are many factors which may contribute to the age associated increased risk of diabetes. Peripheral insulin sensitivity declines with age, and can be attributed to increased obesity, reduced physical activity, and decreased lean muscle mass [2 - 4]. These changes are moderately compensated by increased β-cell function and hyperinsulinemia [5]. However, aging is also associated with a progressive decrease in β-cell function [6 - 8], which may be impacted by decreased incretin levels [9, 10], Sirt1-mediated glucose stimulate insulin release [11], mitochondrial function and ATP content [12, 13], as well as increased oxidative stress [12] and glucolipotoxicity [14]. Moreover, isolated islets demonstrate greater sensitivity to glucose-induced β-cell apoptosis with increasing age [8]. Thus, chronic hyperglycemia and insulin resistance in the setting of aging and declining β-cell function collectively contribute to drive β-cell loss. Human post-mortem samples demonstrate increased rates of apoptosis and reduced β-cell mass in patients with T2D [15]. The inability of β-cells to adequately compensate through increased β-cell mass in the presence of insulin resistance leads to β-cell apoptosis and T2D. Elucidation of the mechanisms of aging that contribute to failed β-cell compensation will be critical for diabetes therapies. Aging is also an important factor in type 1 diabetes (T1D) given that over 50% of T1D cases occur in adults [16], and moreover, with insulin therapy and quality health care, the life expectancy of juveniles with T1D extends well into adulthood [17]. Considering the large adult population with T1D, age associated changes in β-cell function and mass may influenced the design of successful diabetes therapies.

β-CELL PROLIFERATION AND AGING

Pancreatic β-cell mass rapidly increases in young rodents and slowly expands with advancing age [18, 19]. Maintenance and expansion of β-cell mass is dependent on a fine balance between cell birth (self-replication and neogenesis) and cell death (apoptosis). Postnatal β-cell neogenesis is extremely controversial, and while many suggest that β-cells arise from exocrine duct cell differentiation [20 - 22], there is strong evidence to suggest that postnatal β-cell growth exclusively occurs by self-duplication [23 - 27]. Thus, it is critically important to understand the underlying biology governing β-cell proliferation for developing diabetes therapies.

β-cell proliferation in the early postnatal period is ~20% in rodents [18, 28], declining dramatically in adolescence and into young adulthood [18, 19, 28, 29]. This decline stabilizes within the first 100 days of life to ~1-4% per day [18]. Based on this data, it was estimated that the lifespan of a β-cell was between 1 and 3 months [18]. Limited data at older ages prevented conclusive determination of the β-cell replication rate in older rats. Subsequently, Montanya *et al.* [19] measured intra-islet proliferation in rats up to 20 months of age. The authors determined that intra-islet cell proliferation stabilized at ~0.1% using a 6-h BrdU pulse (~0.4% per day) at 7 months of age. Mouse β-cell proliferation was extremely low at 1 year of age, ~0.04% per day, and remained constant at 19-months [28, 30]. Similarly, β-cell apoptosis in metabolically normal rodents is rare and decreases with age [28, 31, 32]. β-cell turnover is governed by the replication refractory period, which prevents cell cycle entry immediately following cell division, thus limiting the frequency a β-cell can divide [24]. This replication refractory period of β-cell turnover is lengthened with age [24, 33]. Collectively, these studies suggest that β-cells are largely post-mitotic and long-lived, contributing to a very slowly expanding adult tissue through self-renewal. Therefore, rodent β-cell turnover is minimal with increasing age.

Human β-cell proliferation also decreases with age [15, 34 - 37]. Meier *et al.* [34] measured β-cell proliferation by Ki67 in 46 tissue samples from children and young adults from 2 weeks to 21 years. In some infants, β-cell proliferation was quite high (~2.5%), and quickly declined with age. Surprisingly, pancreata from several individuals indicated no β-cell proliferation even at an early age (<10 years). Low replication rates were also observed in middle age [35]. By elderly age (~78 years), data from 17 non-diabetic individuals indicated that β-cell proliferation rates had fallen below 0.1% Ki67[+] β-cells [15]. While Ki67[+] expression is a static measurement at the time of pancreas collection, thymidine analog incorporation can provide long-term dynamic measurements. An interesting study examined ten pancreata from individuals that received a thymidine analog (BrdU or IdU) for eight days to four years, as a labeling agent or radio-sensitizer in a clinical trial [37]. Two individuals (18 and 20 years old) contained thymidine positive replicating β-cells (1-2%), however zero replicating β-cells were found in the remaining older individuals (age 31-74), suggesting that β-cell turnover is limited to the first three decades of life, and by supported carbon dating analysis [37]. Cnop *et al.* estimated that human β-cells are long-lived, evident by increased accumulation of lipofusin bodies [36]. In contrast to rodents, low human β-cell turnover and increased longevity appear to contribute to a stable β-cell mass that is established in the first 20-30 years [36, 38]. Species differences and limited β-cell turnover are important considerations for attempted β-cell expansion strategies for human diabetes therapy. These restrictions may limit the utility of attempted β-cell regeneration for diabetes therapies. However, understanding the underlying cellular and molecular changes with advanced age may provide novel targets to stimulate β-cell proliferation.

AGING AND ADAPTIVE β-CELL PROLIFERATION

Pancreatic β-cells demonstrate a remarkable ability to compensate for increasing insulin demands [15, 39 - 41]. A failure to adaptively increase β-cell mass to meet the demand for insulin is thought to lead to T2D [29]. Intriguingly, the majority of obese humans do not develop diabetes, largely due to an adaptive increase in β-cell mass and function [15, 42]. Measurements from pancreatic autopsy samples demonstrated a ~50% increase in β-cell mass in obese non-diabetic samples [15, 38]. Butler and colleagues rule out a significant contribution from β-cell proliferation or hypertrophy [increased cell size], and suggest that neogenesis, observed as an increase in duct associated insulin positive cells, accounts for the increase in β-cell mass [15, 38]. Adaptive β-cell expansion also occurs during pregnancy [43]. The percentage of β-cells increased by ~10% in pregnancy due to hyperplasia (increased cell number) compared to age-matched non-pregnant females. Collectively, these studies suggest that increased insulin demand and reduced insulin sensitivity in obesity and pregnancy can stimulate compensatory β-cell growth.

Although increased insulin resistance with aging correlates with greater β-cell mass [44], it is possible that β-cell expansion in obesity may have occurred at an earlier age when β-cell proliferation is higher. Advanced maternal age is a risk factor for developing gestational diabetes [45], which may suggest a failure to increase β-cell mass in older age, while younger women are capable of adapting. The direct effect of aging on adaptive β-cell growth is difficult to assess in post-mortem samples.

Transplantation of cadaveric human islets into rodents is one approach to gain insight into these phenomena. Gargani *et al.* transplanted human islets into Rag2$^{-/-}$ mice on a high fat diet (HFD) and followed them for up to 12 weeks [46]. Analysis of human grafts suggested increased β-cell proliferation at 4, 6, and 12 weeks. Notably, donor islets were obtained from three lean young adults (16, 16, 22 years old) and one obese adult (41 years old), all of which may have higher basal proliferation rates. Tyrberg *et al.* performed a comprehensive analysis of human β-cell regenerative capacity, transplanting human islets from donors aged ~18 to ~64 years old [47]. Islets were transplanted mice previously rendered hyperglycemic by administration of the selective β-cell toxin alloxan, and found that β-cell proliferation decreased with increasing donor age. Similar findings were found when human donor islets of various ages were transplanted into mice and stimulated with the β-cell mitogen, Exendin-4 (Ex-4) [48]. Thus, evidence from human islet transplantation studies supports the theory of an aged-dependent decline in β-cell regenerative capacity.

Rodents represent a practical model for the analysis of adaptive β-cell growth in aging given the ability to sample from multiple time points with relative ease. In rodents, β-cell regeneration is robust in young mice [30, 49 - 51]. Swenne first interrogated the effects of aging on β-cell replication in response to glucose by culturing isolated islets from fetal and postnatal (1-week, 3-week, 3-4 months) mice [29]. β-cell proliferation was significantly decreased under basal and glucose stimulated conditions in adolescent rat islets, suggesting an age-dependent decrease in β-cell regeneration, which has been confirmed more recently in similar studies [30, 52 - 54]. An elegant study performed by Kushner and colleagues examined three models of β-cell regeneration in a wide range of young and elderly mice (2, 8, 12, 14, 19 months of age) [30]. Partial pancreatectomy in

2-month old mice stimulated a significant increase in β-cell proliferation, which gradually declined until regeneration was abrogated at 14 and 19 months of age. Administration of multiple low doses of the β-cell toxin streptozotocin (STZ), elicited a robust response in 2-month-old mice but no change in β-cell proliferation at 14 months, consistent with other reports [53]. Comparable results were obtained in mice treated for 21 days with Ex-4 [30]. Bhushan and colleagues performed a similar study to test the regenerative capacity of young (2 months) and old (7-8 month old) mice following STZ or Ex-4 treatment, or 8 weeks on a HFD [54]. In each model, old mice failed to increase β-cell proliferation in response to a regenerative stimulus compared to young mice. However, a study by Dor and colleagues suggests that aged β-cells retain their regenerative capacity [55]. β-cell ablation in young mice induced a robust proliferative response, which declined with age (~8% at 5 weeks and <1.5% at 25.5 months). Although, elderly mice (25.5 months) were still capable of increasing β-cell proliferation by up to 5-fold compared to controls, suggesting even very old mice retain some capacity for β-cell regeneration. Similarly, Chen *et al.* found that hyperglycemia equally stimulated β-cell proliferation in transplanted islets of young (3 months) and old (20-24 months) mice [56]. Therefore, adaptive β-cell proliferation decreases with age in rodents, but whether it is diminished or completely abolished with age remains unresolved.

Several important factors must be considered to truly understand the effects of aging on adaptive β-cell proliferation. How we define "old" in rodents is a critical factor when designing studies and interpreting results. The lifespan of a mouse is ~30 months of age [57]. Therefore, an 8-month-old mouse represents about 27% of its lifespan, which is comparable to a 21-year-old human. The study by Tschen *et al.* suggested that adaptive β-cell proliferation was lost in 8-month-old mice in response to STZ, Ex-4, and HFD [54]. In contrast, Rankin *et al.* demonstrated that 12-month-old mice remain responsive to partial pancreatectomy [30], while Dor and colleagues report that 25.5-month-old mice are capable of modest regeneration following β-cell ablation with diphtheria toxin [55]. Consequently, the second factor to consider is the model of β-cell regeneration employed. Partial pancreatectomy appears to be a much stronger stimulus for β-cell proliferation than Ex-4 or HFD. Additionally, hyperglycemia induced by β-cell depletion using dipetheria toxin was more likely to stimulate aged β-cells to proliferate than other β-cell toxins (STZ) [54, 55], possibly related to the method and extent of β-cell damage/loss. Lastly, there are significant differences between mice and humans. Mice have higher metabolic rates, greater basal β-cell proliferation, and may retain some capacity for β-cell regeneration in old age [30, 36, 55, 58]. In contrast, human β-cell proliferation is restricted to the first three decades in patients without diabetes [37], and minimal β-cell regeneration is observed in patients after partial pancreatectomy [59]. Though even in humans some exceptions have been observed, such as increased β-cell proliferation in the pancreas of an 89-year-old patient with recent onset T1D [60]. Therefore, significant consideration must be applied to how aging is modeled in the context of β-cell proliferation and how these observations may translate to humans.

MOLECULAR AND SIGNALING CHANGES IN AGING

Aging is associated with intricate molecular and signaling changes that reduce β-cell turnover and limit adaptive β-cell growth (Fig. 1). A lack of mitogenic signals and/or decreased responsiveness to these signals may underlie the limited capacity for β-cell proliferation in aged mammals. Parabiosis experiments to surgically join young and old mice were performed, thereby creating a common circulation, to determine if old mice lacked the mitogenic signals for β-cell growth, and whether aged β-cells remained responsive to these signals [61]. β-cell proliferation increased in old mice (7-8 months) parabiosed to young mice (1.5 months) suggesting that aged β-cells retain an intrinsic ability to regenerate, but the signals are absent in old mice. While there may be a paucity of circulating β-cell growth factors in old age, reduced sensitivity to these signals may also contribute to the decline in β-cell proliferation. For example, studies have suggested that sensitivity to circulating incretins may be impaired in aged β-cells [9, 62]. Treatment with Ex-4 stimulated β-cell proliferation in young mice, but had no effect on older mice up to 14 months of age [30, 54]. Similarly, platelet derived growth factor (PDGF) receptor expression decreases with age [63], and PDGF treatment of 12-month-old mouse or adult human islets *in vitro* had no effect on β-cell proliferation as expected. Hepatic expression of betatrophin, a recently proposed β-cell mitogen [40], increases with age although β-cell proliferation declines in old age [64]. Therefore, a combination of absence of growth factors and reduced β-cell sensitivity to these signals may limit β-cell proliferation with aging.

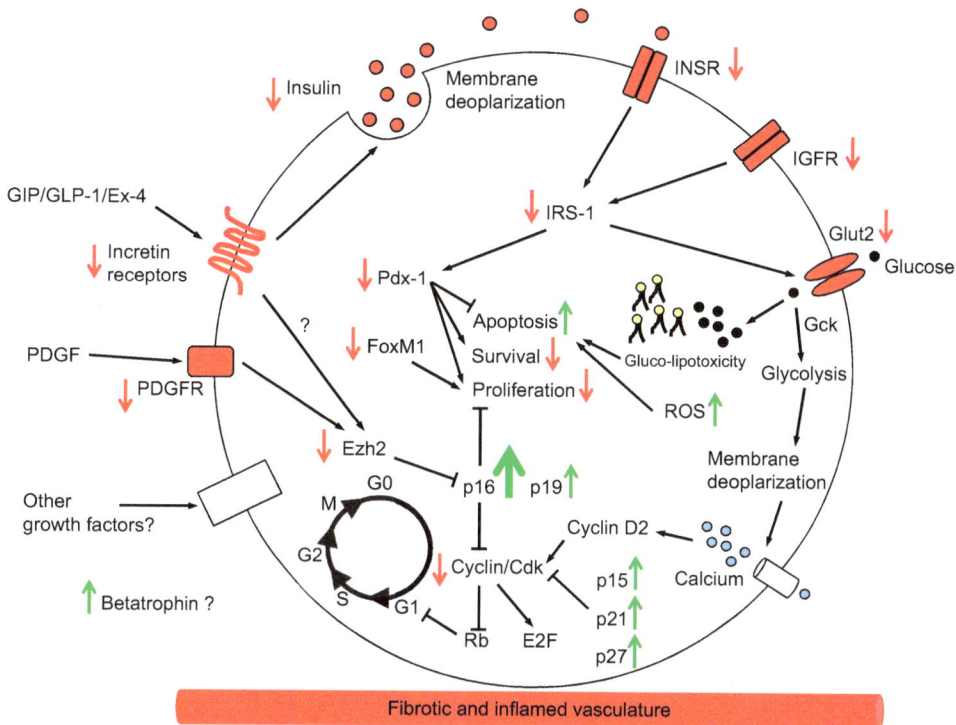

Fig. (1). β-cell specific changes in molecular signals associated with aging. A schematic representation of the changes that occur in aging to decrease β-cell proliferation and survival, with increased cell death. Receptor expression of several growth factors is decreased with age; the platelet derived growth factor (PDGF) receptor (PDGFR), incretin receptors that bind glucose-dependent insulinotropic peptide (GIP), glucagon-like peptide-1 (GLP-1), or exendin-4 (Ex-4), insulin receptor (INSR), and insulin-like growth factor receptor (IGFR). Loss of signaling through these pathways may correspond with a decrease in Ezh2 expression through the PDGFR and possibly Ex-4, which will impact epigenetic regulation of p16/p19. A decrease in IRS-1, INSR, IGFR signaling contributes to a decrease in Pdx-1 expression, which regulates proliferation, survival, and apoptosis. Betatrophin expression increases with age, but its receptor is unknown and the direct effects on β-cells are unclear. Glucose enters the β-cell through Glut2 and progresses through glycolysis *via* Gck, which subsequently induces membrane depolarization and calcium entry into the cell. Glucose-induced β-cell proliferation is dependent on intracellular calcium; however, Glut2 expression decreases with age and may limit glucose induced β-cell proliferation. On the other hand, excess glucose and lipid accumulation can lead to gluco-lipotoxicity, reactive oxygen species (ROS) and β-cell death. Cell cycle inhibitors (p15, p16, 19, p21, p27) increase with age and inhibit activity of Cyclin dependent kinase (Cdk) and Cylin (largely D-type Cyclins) complexes, preventing cell cycle progression, thus reducing β-cell proliferation. Lastly, the islet vasculature becomes fibrotic and inflamed with age, which can limit the delivery of growth factors and nutrients required to induce β-cell proliferation and survival. (Red – decreased, Green – increased).

Reduced sensitivity to mitogenic signals may also result from impaired downstream signaling in the aged β-cell. FoxM1 is highly expressed in proliferating cells, and adaptive β-cell proliferation in insulin resistant rats was dependent on FoxM1; however, FoxM1 expression declines with age [41, 65, 66]. Insulin is known to have strong growth promoting effects on β-cells [67 - 69]. Several genes relating to the insulin signaling pathway, including the insulin receptor, insulin-like growth factor 1 receptor (IGF-1R), insulin receptor substrate (IRS) -1, and pancreatic and duodenal homoebox 1 (Pdx-1) are decreased with age in rodents and may limit the growth promoting effects of insulin [70, 71]. By contrast, IRS-2 and Akt expression levels are maintained in 11-month-old mice [70]. Notably, there is also an age-related decline in insulin mRNA expression and insulin secretion [6 - 8, 70]. Within the insulin signaling pathway, a reduction of Pdx-1 expression alone may account for significant deficits in β-cell turnover because of its critical role to promote β-cell survival and reduce apoptosis [72 - 74]. Pdx-1 expression also decreases with age in human β-cells [35]. Significant changes in β-cell signaling molecules are greatly reduced with age, which may prevent signal transduction of β-cell mitogenic signals.

Glucose is a robust stimulus for β-cell proliferation in young rodents/mice [75, 76]. In aged mammals, hyperglycemia induced β-cell proliferation was equivalent in young (3-month) and old (20-24 month) mice [56]. Hyperglycemia resulting from β-cell ablation also stimulated β-cell proliferation in mice >20 months of age [55]. Dor and colleagues suggest that glucose regulates β-cell proliferation through a pathway involving glucokinase, which senses glucose, glycolysis, membrane depolarization, and calcium-dependent upregulation of Cyclin D2 [76, 77]. Presumably these mechanisms remain intact in aged mice, specifically facilitating glucose-induced β-cell proliferation. The direct link between Cyclin D2 and glucose sensing is unclear, and paradoxically, Cyclin D2 and Glut2 expression decreases with age in normoglycemic mice [70, 71, 78], although it is possible expression is restored during hyperglycemia. Assefa *et al.* [79] reported that glucose had no effect on β-cell growth *in vitro* examining islets isolated from 40-week-old rats. In humans, aging was correlated with an increased sensitivity of β-cells to glucose-induced apoptosis and decreased β-cell proliferation [8], which may be attributed to a decrease in Pdx-1 expression with age [8, 35]. Therefore, the effect of aging on glucose-induced β-cell proliferation appears to be complicated, and the signals that regulate this process require further investigation in aged mice.

Reduced growth factor and glucose induced β-cell proliferation may be due to poor islet vascularization in aged mice. β-cell growth and regeneration are dependent on the supportive islet microvasculature [80, 81]. A recent study found that the islet vasculature from 18-month old mouse islets and adult human islets was associated with chronic inflammation and fibrosis [82]. Vascular aging was reversed when aged mouse islets were transplanted into the anterior chamber of the eye (an easily accessible and highly vascularized region) in young recipients. Subsequently, β-cell proliferation was increased in aged islet grafts. Therefore, age-restricted β-cell proliferation may be influenced by islet vascularization, in addition to systemic growth factors and intrinsic β-cell signaling.

Growth factor and intracellular signaling pathways converge on molecular signals that govern cell cycle entry and progression. These cell cycle regulators are strongly influenced by age. An important step in cell cycle progression is the G_1/S transition, wherein Cyclin D and cyclin dependent kinase (Cdk) complexes phosphorylate and inactive retinoblastoma protein, allowing E2F induced gene transcription of cell cycle genes. β-cell expression of Cdk2, Cyclin D1-3, and Cyclin E1-2 was reduced in mice ≥11 months of age, compared to young mice [30, 70]. Partial pancreatectomy stimulated islet expression of several Cyclins (A2, D1, E1, E2) in young mice, but these genes were all downregulated or unchanged in 14-month-old partial pancreatectomized mice [83]. Cyclin dependent kinase inhibitors, such as the INK/ARF (p15, p16, p18, p19) family of proteins, regulate Cyclin D/Cdk complexes. The INK4a/Arf locus encodes for the tumor suppressors, p16[INK4a] and p19[Arf], which are biomarkers of aging in many tissues [84]. Both rodent and human islets have increased expression of p16 with age, correlating with the decline in β-cell proliferation [30, 53, 54, 85]. β-cell overexpression of p16 in young mice decreased β-cell proliferation to rates observed in older mice, while p16 deletion rescued the β-cell proliferative capacity in aged mice [53]. HFD or Ex-4 treatment both decreased p16 expression in young mice, but had no effect at 7-8 months [54]. These observations strongly implicate p16 in the age dependent decline in β-cell proliferation. In addition to p16, several other cell cycle inhibitors, such as p15, p21, and p27 have been noted to increase with age [30]. Consistent with an increase in cycle cell inhibitors, the replication refractory period of β-cell turnover is lengthened with age [24, 33]. Altered cell cycle gene expression and subsequent kinetics may be primarily responsible for the reduction in β-cell proliferation in aging.

Cell cycle genes are subject to epigenetic regulation, which are associated with age-specific changes in pancreatic β-cells (Fig. **2**). Polycomb group (PcG) protein complexes induce epigenetic silencing of genes through chromatin remodeling, and antagonize Trithorax group (TxG) proteins, which activate gene expression. The Ink4a/Arf locus, which expresses p16 and p19, is highly regulated by these regulatory complexes. Bmi-1 is part of the Polycomb-repressive complex 1 (PRC1) with ubiquitin E3 ligase activity, which targets histone H2a for ubiquitination [86]. PRC2 contains Ezh2, with H3K27 methyltransferase activity [87 - 90]. PRC2 trimethylation of H3K27 recruits PRC1 and ubiquitination of H2A to suppress gene transcription [85, 91]. Opposing PcG action is TxG, which contains Mll with H3K4 methyltransferase activity. Bmi-1 and Ezh2 expression decrease with age, leading to reduced inhibition of the Ink4a/Arf locus, thereby increasing expression of p16 and p19 [85, 92]. Similarly, loss of Bmi-1 or Ezh2 in young mice results in increased p16 expression associated with specific changes in epigenetic marks [54, 92]. Bmi-1 knockout results in decreased H2a ubiquitination and increased H3K9 acetylation by ChIP analysis [85]. siRNA or β-cell specific deletion of Ezh2 decreases H3K27 trimethylation and increases H3K4 trimethylation [85, 92]. This remodeling of the chromatin will favour gene transcription of the Ink4a/Arf locus and inhibition of β-cell proliferation. Moreover, Ezh2 siRNA knockdown decreased Bmi-1 binding at the Ink4a/Arf locus, and decreased H2A ubiquitination, supporting a model of PRC1/Bmi-1 recruitment through PRC2/Ezh2 trimethylation of H3K27 [85, 91]. Loss of Bmi-1 or Ezh2 expression in young mice significantly decreases β-cell proliferation and mass, which can be rescued by deletion of the Ink4a/Arf locus [85, 92]. β-cell overexpression of Ezh2 increased β-cell mass and decreased p16 expression in young mice [93]. In summary, p16 expression increases with aging due to reduced PcG suppression and is associated with decreased β-cell proliferation.

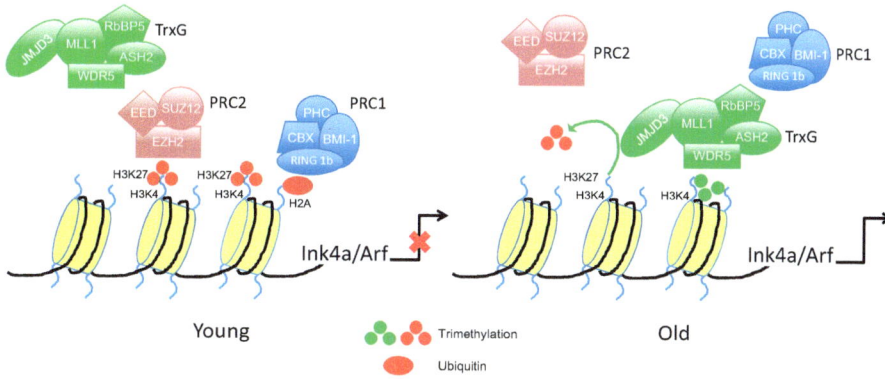

Fig. (2). Epigenetic regulation of the Ink4a/Arf locus in young and old mammals. The polycomb-repressive complexes (PRC) suppress expression of p16 and p19 from the Ink4a/Arf locus in young mammals. PRC1 is composed of B cell-specific Moloney murine leukemia virus integration site 1 (Bmi-1), E3 ubiquitin-protein ligase (Ring 1b), Chromobox Homolog (CBX), and Polyhomeotic homolog (PHC). PRC1 is recruited to the chromatin by PRC2 for the ubiquitination (red oval) of histone H2A. PRC2 is composed of Enhancer of Zeste 2 (Ezh2), Embryonic ectoderm development (Eed), and Suppressor of Zeste 12 protein homolog (Suz12), which possess histone methyltransferase activity and trimethylates (red circles) histone H3 at lysine 27 (H3K27). With advancing age, decreased expression of Bmi-1 and Ezh2, with increased activity of the Trithorax Group (TrxG) proteins results in increased gene transcription. The TrxG protein complex is composed of myeloid/lymphoid or mixed-lineage leukemia (Mll1), WD repeat-containing protein 5 (Wdr5), absent, small or homeotic discs 2 (Ash2), retinoblastoma-binding protein 5 (RbBP5) to trimethylate (green circles) histone H3 at lysine 4 (H3K4), which limits recruitment of PRC2. Mll-1 also recruits Jumonji domain containing 3 (JmjD3), a demethylase, to remove the methyl groups from H3K27, and results in increased p16 and p19 expression. (Red – repressive epigenetic marks; Green – active epigenetic marks).

Conversely, TrxG proteins induce chromatin remodeling to activate gene transcription. Mll associates with the demethylase, JmjD3, in the TxG complex. JmjD3 increases with age in human islets, and binds to Ink4a to demethylate H3K27 [92, 93]. Knockdown of Mll or Jmjd3 in islets overexpressing Ezh2, decreased H3K4me3, and increased Ezh2 recruitment, H3K27me3, and β-cell proliferation [93]. Furthermore, Mll knockdown results in decreased p16 protein expression. Therefore, TrxG activity at the Ink4a/Arf locus may result in increased p16 levels and decreased β-cell proliferation in aging.

β-cell regeneration reverses these epigenetic marks in young mice to suppress p16 expression and increase β-cell proliferation. Mice placed on HFD for 8 weeks or treated for 7 days with Ex-4 had increased Bmi-1 and reduced p16 expression, thereby increasing β-cell proliferation and mass [54]. However, a germline deletion of Bmi-1 abrogated the effects of Ex-4 on β-cell proliferation and mass [54]. STZ treatment increased Ezh2 expression with a corresponding downregulation of p16 [92]. β-cell specific loss Ezh2 significantly reduced β-cell regeneration following STZ. These findings demonstrate the importance of repressing p16 expression to facilitate β-cell regeneration. In old mice, β-cell regeneration is limited or absent [30, 54]. Treatment of adult mice with a HFD or Ex-4 had no effect on p16 expression, which remained elevated compared to controls, with significantly low levels of Bmi-1 expression. Thus, old mice are unable to upregulate Bmi-1 and suppress p16 expression to allow for compensatory β-cell proliferation. These studies support a major role of p16 in limiting β-cell regeneration in old age.

In summary, aging significantly diminishes the genetic and molecular signaling pathways that regulate β-cell proliferation. The severe decline in β-cell proliferation may be attributed to several different intervening points in a complex network. First, the initiation of the proliferative response may be affected by loss of circulating or local mitogenic factors and their corresponding receptors on β-cells. Second, the transduction of the mitogenic signal may be lost due to decreased expression of intracellular mediators such as kinases and transcription factors. Finally, cell cycle progression may be slowed or shunted into senescence by genetic and epigenetic changes that increase cell cycle inhibitors and reduce promoters of cell cycle progression. Clearly aging has a tremendous impact on multiple levels of a complex regulatory network of β-cell proliferation.

SIGNIFICANCE FOR TARGETED β-CELL REGENERATION

A substantial loss of functional β-cell mass occurs in both T1D and T2D. Regeneration of endogenous β-cells is a promising therapeutic approach to cure and prevent diabetes, however there are several obstacles limiting successful β-cell regeneration, including aging. β-cell proliferation declines with age, reducing the capacity for β-cell regeneration. However, some studies suggest that even very old mouse β-cells can be forced to enter cell cycle, giving rise to the possibility that even human β-cells may be capable of cell proliferation under the right conditions. What these conditions are, and whether there is a stimulus or factor strong enough to overcome the age-induced restrictions on β-cell proliferation is currently unknown. Presumably, such a factor would require a responsive β-cell capable of signal transduction that would reverse the epigenetic changes in cell cycle inhibitors, freeing β-cells to divide. Long acting forms of Ex-4 are FDA approved for the treatment of T2D, and also in clinical trials for T1D [94 - 96]. In addition to the ability of Ex-4 to improve β-cell function, there is hope that patients will also benefit from increased β-cell mass, as is observed in mice [54]. Small molecule inhibitors are being developed to target epigenetic regulators of gene transcription [97, 98]. Most of these drugs aim to suppress proliferation for cancer therapy; however, one drug inhibits Mll, a component of TrxG proteins that modify chromatin to activate gene transcription of the cell cycle inhibitor p16. Potentially, small molecules that inhibit activators of p16, thereby repressing p16 expression, would allow for β-cell proliferation. In conclusion, aging induces significant molecular changes that reduce the capacity for β-cell proliferation. On-going research will aim to improve our understanding of the effects of aging on β-cell proliferation and elucidate new drug targets for diabetes therapy.

CONFLICT OF INTEREST

The author confirms that author has no conflict of interest to declare for this publication.

ACKNOWLEDGEMENTS

Declared none.

REFERENCES

[1]   Cowie CC, Rust KF, Ford ES, *et al.* Full accounting of diabetes and pre-diabetes in the U.S. population in 1988-1994 and 2005-2006. Diabetes Care 2009; 32(2): 287-94.
[http://dx.doi.org/10.2337/dc08-1296] [PMID: 19017771]

[2]   Rowe JW, Minaker KL, Pallotta JA, Flier JS. Characterization of the insulin resistance of aging. J Clin Invest 1983; 71(6): 1581-7.
[http://dx.doi.org/10.1172/JCI110914] [PMID: 6345585]

[3]   Sheen AJ. Diabetes mellitus in the elderly: insulin resistance and/or impaired insulin secretion? Diabetes Metab 2005; 31: 5S27-34.

[4]   Gunasekaran U, Gannon M. Type 2 diabetes and the aging pancreatic β cell. Aging (Albany, NY) 2011; 3(6): 565-75.

[http://dx.doi.org/10.18632/aging.100350] [PMID: 21765202]

[5]   Prentki M, Nolan CJ. Islet β-cell failure in type 2 diabetes. J Clin Invest 2006; 116(7): 1802-12.
        [http://dx.doi.org/10.1172/JCI29103] [PMID: 16823478]

[6]   Muzumdar R, Ma X, Atzmon G, Vuguin P, Yang X, Barzilai N. Decrease in glucose-stimulated
        insulin secretion with aging is independent of insulin action. Diabetes 2004; 53(2): 441-6.
        [http://dx.doi.org/10.2337/diabetes.53.2.441] [PMID: 14747296]

[7]   Wang SY, Halban PA, Rowe JW. Effects of aging on insulin synthesis and secretion. Differential
        effects on preproinsulin messenger RNA levels, proinsulin biosynthesis, and secretion of newly made
        and preformed insulin in the rat. J Clin Invest 1988; 81(1): 176-84.
        [http://dx.doi.org/10.1172/JCI113291] [PMID: 3275693]

[8]   Maedler K, Schumann DM, Schulthess F, *et al.* Aging correlates with decreased β-cell proliferative
        capacity and enhanced sensitivity to apoptosis: a potential role for Fas and pancreatic duodenal
        homeobox-1. Diabetes 2006; 55(9): 2455-62.
        [http://dx.doi.org/10.2337/db05-1586] [PMID: 16936193]

[9]   Elahi D, Andersen DK, Muller DC, Tobin JD, Brown JC, Andres R. The enteric enhancement of
        glucose-stimulated insulin release. The role of GIP in aging, obesity, and non-insulin-dependent
        diabetes mellitus. Diabetes 1984; 33(10): 950-7.
        [http://dx.doi.org/10.2337/diab.33.10.950] [PMID: 6383904]

[10]   Meneilly GS, Ryan AS, Minaker KL, Elahi D. The effect of age and glycemic level on the response of
        the β-cell to glucose-dependent insulinotropic polypeptide and peripheral tissue sensitivity to
        endogenously released insulin. J Clin Endocrinol Metab 1998; 83(8): 2925-32.
        [PMID: 9709971]

[11]   Ramsey KM, Mills KF, Satoh A, Imai S. Age-associated loss of Sirt1-mediated enhancement of
        glucose-stimulated insulin secretion in β cell-specific Sirt1-overexpressing (BESTO) mice. Aging Cell
        2008; 7(1): 78-88.
        [http://dx.doi.org/10.1111/j.1474-9726.2007.00355.x] [PMID: 18005249]

[12]   Cooksey RC, Jouihan HA, Ajioka RS, *et al.* Oxidative stress, β-cell apoptosis, and decreased insulin
        secretory capacity in mouse models of hemochromatosis. Endocrinology 2004; 145(11): 5305-12.
        [http://dx.doi.org/10.1210/en.2004-0392] [PMID: 15308612]

[13]   Anello M, Lupi R, Spampinato D, *et al.* Functional and morphological alterations of mitochondria in
        pancreatic β cells from type 2 diabetic patients. Diabetologia 2005; 48(2): 282-9.
        [http://dx.doi.org/10.1007/s00125-004-1627-9] [PMID: 15654602]

[14]   Fontés G, Zarrouki B, Hagman DK, *et al.* Glucolipotoxicity age-dependently impairs β cell function in
        rats despite a marked increase in β cell mass. Diabetologia 2010; 53(11): 2369-79.
        [http://dx.doi.org/10.1007/s00125-010-1850-5] [PMID: 20628728]

[15]   Butler AE, Janson J, Bonner-Weir S, Ritzel R, Rizza RA, Butler PC. β-cell deficit and increased β-cell
        apoptosis in humans with type 2 diabetes. Diabetes 2003; 52(1): 102-10.

[16]   Wang X, He Z, Ghosh S. Investigation of the age-at-onset heterogeneity in type 1 diabetes through
        mathematical modeling. Math Biosci 2006; 203(1): 79-99.
        [http://dx.doi.org/10.1016/j.mbs.2006.03.021] [PMID: 16723139]

[17]   Gale EA. How to survive diabetes. Diabetologia 2009; 52(4): 559-67.
[http://dx.doi.org/10.1007/s00125-009-1275-1] [PMID: 19198799]

[18]   Finegood DT, Scaglia L, Bonner Weir S. Dynamics of β-cell mass in the growing rat pancreas. Estimation with a simple mathematical model. Diabetes 1995; 44(3): 249-56.

[19]   Montanya E, Nacher V, Biarnés M, Soler J. Linear correlation between β-cell mass and body weight throughout the lifespan in Lewis rats: role of β-cell hyperplasia and hypertrophy. Diabetes 2000; 49(8): 1341-6.
[http://dx.doi.org/10.2337/diabetes.49.8.1341] [PMID: 10923635]

[20]   Bonner-Weir S, Weir GC. New sources of pancreatic β-cells. Nat Biotechnol 2005; 23(7): 857-61.
[http://dx.doi.org/10.1038/nbt1115] [PMID: 16003374]

[21]   Inada A, Nienaber C, Katsuta H, *et al.* Carbonic anhydrase II-positive pancreatic cells are progenitors for both endocrine and exocrine pancreas after birth. Proc Natl Acad Sci USA 2008; 105(50): 19915-9.
[http://dx.doi.org/10.1073/pnas.0805803105] [PMID: 19052237]

[22]   Xu X, DHoker J, Stangé G, *et al.* β cells can be generated from endogenous progenitors in injured adult mouse pancreas. Cell 2008; 132(2): 197-207.
[http://dx.doi.org/10.1016/j.cell.2007.12.015] [PMID: 18243096]

[23]   Dor Y, Brown J, Martinez OI, Melton DA. Adult pancreatic β-cells are formed by self-duplication rather than stem-cell differentiation. Nature 2004; 429(6987): 41-6.
[http://dx.doi.org/10.1038/nature02520] [PMID: 15129273]

[24]   Teta M, Rankin MM, Long SY, Stein GM, Kushner JA. Growth and regeneration of adult β cells does not involve specialized progenitors. Dev Cell 2007; 12(5): 817-26.
[http://dx.doi.org/10.1016/j.devcel.2007.04.011] [PMID: 17488631]

[25]   Brennand K, Huangfu D, Melton D. All β cells contribute equally to islet growth and maintenance. PLoS Biol 2007; 5(7): e163.
[http://dx.doi.org/10.1371/journal.pbio.0050163] [PMID: 17535113]

[26]   Kopp JL, Dubois CL, Schaffer AE, *et al.* Sox9+ ductal cells are multipotent progenitors throughout development but do not produce new endocrine cells in the normal or injured adult pancreas. Development 2011; 138(4): 653-65.
[http://dx.doi.org/10.1242/dev.056499] [PMID: 21266405]

[27]   Rankin MM, Wilbur CJ, Rak K, Shields EJ, Granger A, Kushner JA. β-Cells are not generated in pancreatic duct ligation-induced injury in adult mice. Diabetes 2013; 62(5): 1634-45.
[http://dx.doi.org/10.2337/db12-0848] [PMID: 23349489]

[28]   Teta M, Long SY, Wartschow LM, Rankin MM, Kushner JA. Very slow turnover of β-cells in aged adult mice. Diabetes 2005; 54(9): 2557-67.
[http://dx.doi.org/10.2337/diabetes.54.9.2557] [PMID: 16123343]

[29]   Swenne I. Effects of aging on the regenerative capacity of the pancreatic B-cell of the rat. Diabetes 1983; 32(1): 14-9.

[30]   Rankin MM, Kushner JA. Adaptive β-cell proliferation is severely restricted with advanced age. Diabetes 2009; 58(6): 1365-72.
[http://dx.doi.org/10.2337/db08-1198] [PMID: 19265026]

[31] Scaglia L, Cahill CJ, Finegood DT, Bonner-Weir S. Apoptosis participates in the remodeling of the endocrine pancreas in the neonatal rat. Endocrinology 1997; 138(4): 1736-41.
[http://dx.doi.org/10.1210/en.138.4.1736]

[32] Petrik J, Reusens B, Arany E, *et al.* A low protein diet alters the balance of islet cell replication and apoptosis in the fetal and neonatal rat and is associated with a reduced pancreatic expression of insulin-like growth factor-II Endocrinology 1999; 140(10): 4861-73.
[http://dx.doi.org/10.1210/en.140.10.4861]

[33] Salpeter SJ, Klein AM, Huangfu D, Grimsby J, Dor Y. Glucose and aging control the quiescence period that follows pancreatic β cell replication. Development 2010; 137(19): 3205-13.
[http://dx.doi.org/10.1242/dev.054304] [PMID: 20823063]

[34] Meier JJ, Butler AE, Saisho Y, *et al.* β-cell replication is the primary mechanism subserving the postnatal expansion of β-cell mass in humans. Diabetes 2008; 57(6): 1584-94.
[http://dx.doi.org/10.2337/db07-1369] [PMID: 18334605]

[35] Reers C, Erbel S, Esposito I, *et al.* Impaired islet turnover in human donor pancreata with aging. Eur J Endocrinol 2009; 160(2): 185-91.
[http://dx.doi.org/10.1530/EJE-08-0596] [PMID: 19004984]

[36] Cnop M, Hughes SJ, Igoillo-Esteve M, *et al.* The long lifespan and low turnover of human islet β cells estimated by mathematical modelling of lipofuscin accumulation. Diabetologia 2010; 53(2): 321-30.
[http://dx.doi.org/10.1007/s00125-009-1562-x] [PMID: 19855953]

[37] Perl S KJ, Buchholz BA, Meeker AK, *et al.* Significant human β-cell turnover is limited to the first three decades of life as determined by In-vivo thymidine analog incorporation and radiocarbon dating J Clin Endocrinol Metab 2010; 95(10): E234-9.

[38] Saisho Y, Butler AE, Manesso E, Elashoff D, Rizza RA, Butler PC. β-cell mass and turnover in humans: effects of obesity and aging. Diabetes Care 2013; 36(1): 111-7.
[http://dx.doi.org/10.2337/dc12-0421] [PMID: 22875233]

[39] El Ouaamari A, Kawamori D, Dirice E, *et al.* Liver-derived systemic factors drive β cell hyperplasia in insulin-resistant states. Cell Reports 2013; 3(2): 401-10.
[http://dx.doi.org/10.1016/j.celrep.2013.01.007] [PMID: 23375376]

[40] Yi P, Park JS, Melton DA. Betatrophin: a hormone that controls pancreatic β cell proliferation. Cell 2013; 153(4): 747-58.
[http://dx.doi.org/10.1016/j.cell.2013.04.008] [PMID: 23623304]

[41] Zarrouki B, Benterki I, Fontés G, *et al.* Epidermal growth factor receptor signaling promotes pancreatic β-cell proliferation in response to nutrient excess in rats through mTOR and FOXM1. Diabetes 2014; 63(3): 982-93.
[http://dx.doi.org/10.2337/db13-0425] [PMID: 24194502]

[42] Polonsky KS. Dynamics of insulin secretion in obesity and diabetes. Int J Obs Relat Metab Disord 2000; 24 (Suppl 2): S29-31.
[http://dx.doi.org/10.1038/sj.ijo.0801273]

[43] Van Assche FA, Aerts L, De Prins F. A morphological study of the endocrine pancreas in human pregnancy. Br J Obstet Gynaecol 1978; 85(11): 818-20.

[http://dx.doi.org/10.1111/j.1471-0528.1978.tb15835.x] [PMID: 363135]

[44] Matveyenko AV, Veldhuis JD, Butler PC. Adaptations in pulsatile insulin secretion, hepatic insulin clearance, and β-cell mass to age-related insulin resistance in rats. Am J Physiol Endocrinol Metab 2008; 295(4): E832-41.
[http://dx.doi.org/10.1152/ajpendo.90451.2008] [PMID: 18664594]

[45] Solomon CG, Willett WC, Carey VJ, *et al.* A prospective study of pregravid determinants of gestational diabetes mellitus. JAMA 1997; 278(13): 1078-83.
[http://dx.doi.org/10.1001/jama.1997.03550130052036] [PMID: 9315766]

[46] Gargani S, Thévenet J, Yuan JE, *et al.* Adaptive changes of human islets to an obesogenic environment in the mouse. Diabetologia 2013; 56(2): 350-8.
[http://dx.doi.org/10.1007/s00125-012-2775-y] [PMID: 23192693]

[47] Tyrberg B, Eizirik DL, Hellerström C, Pipeleers DG, Andersson A. Human pancreatic β-cell deoxyribonucleic acid-synthesis in islet grafts decreases with increasing organ donor age but increases in response to glucose stimulation *In vitro*. Endocrinology 1996; 137(12): 5694-9.
[PMID: 8940401]

[48] Tian L, Gao J, Weng G, *et al.* Comparison of exendin-4 on β-cell replication in mouse and human islet grafts. Transpl Int 2011; 24(8): 856-64.
[http://dx.doi.org/10.1111/j.1432-2277.2011.01275.x] [PMID: 21627696]

[49] Hess D, Li L, Martin M, *et al.* Bone marrow-derived stem cells initiate pancreatic regeneration. Nat Biotechnol 2003; 21(7): 763-70.
[http://dx.doi.org/10.1038/nbt841] [PMID: 12819790]

[50] Nir T, Melton DA, Dor Y. Recovery from diabetes in mice by β cell regeneration. J Clin Invest 2007; 117(9): 2553-61.
[http://dx.doi.org/10.1172/JCI32959] [PMID: 17786244]

[51] Cox AR, Gottheil SK, Arany EJ, Hill DJ. The effects of low protein during gestation on mouse pancreatic development and β cell regeneration. Pediatr Res 2010; 68(1): 16-22.
[http://dx.doi.org/10.1203/PDR.0b013e3181e17c90] [PMID: 20386490]

[52] Tanigawa K, Nakamura S, Kawaguchi M, Xu G, Kin S, Tamura K. Effect of aging on B-cell function and replication in rat pancreas after 90% pancreatectomy. Pancreas 1997; 15(1): 53-9.
[http://dx.doi.org/10.1097/00006676-199707000-00008] [PMID: 9211493]

[53] Krishnamurthy J, Ramsey MR, Ligon KL, *et al.* p16INK4a induces an age-dependent decline in islet regenerative potential. Nature 2006; 443(7110): 453-7.
[http://dx.doi.org/10.1038/nature05092] [PMID: 16957737]

[54] Tschen SI, Dhawan S, Gurlo T, Bhushan A. Age-dependent decline in β-cell proliferation restricts the capacity of β-cell regeneration in mice. Diabetes 2009; 58(6): 1312-20.
[http://dx.doi.org/10.2337/db08-1651] [PMID: 19228811]

[55] Stolovich-Rain M, Hija A, Grimsby J, Glaser B, Dor Y. Pancreatic β cells in very old mice retain capacity for compensatory proliferation. J Biol Chem 2012; 287(33): 27407-14.
[http://dx.doi.org/10.1074/jbc.M112.350736] [PMID: 22740691]

[56] Chen X, Zhang X, Chen F, Larson CS, Wang LJ, Kaufman DB. Comparative study of regenerative

potential of β cells from young and aged donor mice using a novel islet transplantation model. Transplantation 2009; 88(4): 496-503.
[http://dx.doi.org/10.1097/TP.0b013e3181b0d2ee] [PMID: 19696632]

[57]   Bluher M, Kahn BB, Kahn CR. Extended longevity in mice lacking the insulin receptor in adipose tissue. Science 2003; 299(5606): 572-4.
[http://dx.doi.org/10.1126/science.1078223]

[58]   Demetrius L. Of mice and men. When it comes to studying ageing and the means to slow it down, mice are not just small humans. EMBO Rep 2005; 6(Spec No): S39-44.
[http://dx.doi.org/10.1038/sj.embor.7400422] [PMID: 15995660]

[59]   Menge BA, Tannapfel A, Belyaev O, *et al.* Partial pancreatectomy in adult humans does not provoke β-cell regeneration. Diabetes 2008; 57(1): 142-9.
[http://dx.doi.org/10.2337/db07-1294] [PMID: 17959931]

[60]   Meier JJ, Lin JC, Butler AE, Galasso R, Martinez DS, Butler PC. Direct evidence of attempted β cell regeneration in an 89-year-old patient with recent-onset type 1 diabetes. Diabetologia 2006; 49(8): 1838-44.
[http://dx.doi.org/10.1007/s00125-006-0308-2] [PMID: 16802132]

[61]   Salpeter SJ, Khalaileh A, Weinberg-Corem N, Ziv O, Glaser B, Dor Y. Systemic regulation of the age-related decline of pancreatic β-cell replication. Diabetes 2013; 62(8): 2843-8.
[http://dx.doi.org/10.2337/db13-0160] [PMID: 23630298]

[62]   Brown JE. The ageing pancreas. British J Diabetes Vascu Dis 2012; 12(3): 141-5.
[http://dx.doi.org/10.1177/1474651412446713]

[63]   Chen H, Gu X, Liu Y, *et al.* PDGF signalling controls age-dependent proliferation in pancreatic β-cells. Nature 2011; 478(7369): 349-55.
[http://dx.doi.org/10.1038/nature10502] [PMID: 21993628]

[64]   Cox AR, Lam CJ, Bonnyman CW, Chavez J, Rios JS, Kushner JA. Angiopoietin-like protein 8 (ANGPTL8)/betatrophin overexpression does not increase β cell proliferation in mice. Diabetologia 2015; 58(7): 1523-31.
[http://dx.doi.org/10.1007/s00125-015-3590-z] [PMID: 25917759]

[65]   Krupczak-Hollis K, Wang X, Dennewitz MB, Costa RH. Growth hormone stimulates proliferation of old-aged regenerating liver through forkhead box m1b. Hepatology 2003; 38(6): 1552-62.
[http://dx.doi.org/10.1016/j.hep.2003.08.052] [PMID: 14647066]

[66]   Zhang H, Ackermann AM, Gusarova GA, *et al.* The FoxM1 transcription factor is required to maintain pancreatic β-cell mass. Mol Endocrinol 2006; 20(8): 1853-66.
[http://dx.doi.org/10.1210/me.2006-0056] [PMID: 16556734]

[67]   Kulkarni RN, Bruning JC, Winnay JN, Postic C, Magnuson MA, Kahn CR. Tissue specific knockout of the insulin receptor in pancreatic β-cells creates an insulin secretory defect similar to that in type 2 diabetes. Cells 1999; 96(3): 329-9.

[68]   Okada T, Liew CW, Hu J, *et al.* Insulin receptors in β-cells are critical for islet compensatory growth response to insulin resistance. Proc Natl Acad Sci USA 2007; 104(21): 8977-82.
[http://dx.doi.org/10.1073/pnas.0608703104] [PMID: 17416680]

[69]   Folli F, Okada T, Perego C, *et al.* Altered insulin receptor signalling and β-cell cycle dynamics in type 2 diabetes mellitus. PLoS One 2011; 6(11): e28050.
[http://dx.doi.org/10.1371/journal.pone.0028050] [PMID: 22140505]

[70]   Hinault C, Hu J, Maier BF, Mirmira RG, Kulkarni RN. Differential expression of cell cycle proteins during ageing of pancreatic islet cells. Diabetes Obes Metab 2008; 10 (Suppl. 4): 136-46.
[http://dx.doi.org/10.1111/j.1463-1326.2008.00947.x] [PMID: 18834441]

[71]   Ihm SH, Moon HJ, Kang JG, *et al.* Effect of aging on insulin secretory function and expression of β cell function-related genes of islets. Diabetes Res Clin Pract 2007; 77 (Suppl. 1): S150-4.
[http://dx.doi.org/10.1016/j.diabres.2007.01.049] [PMID: 17467845]

[72]   Stoffers DA, Ferrer J, Clarke WL, Habener JF. Early-onset type-II diabetes mellitus (MODY4) linked to IPF1. Nat Genet 1997; 17(2): 138-9.
[http://dx.doi.org/10.1038/ng1097-138] [PMID: 9326926]

[73]   Ahlgren U, Jonsson J, Jonsson L, Simu K, Edlund H. β-cell-specific inactivation of the mouse Ipf1/Pdx1 gene results in loss of the beta-cell phenotype and maturity onset diabetes. Genes Dev 1998; 12(12): 1763-8.

[74]   Johnson JD, Ahmed NT, Luciani DS, *et al.* Increased islet apoptosis in Pdx1+/- mice. J Clin Invest 2003; 111(8): 1147-60.

[75]   Alonso LC, Yokoe T, Zhang P, *et al.* Glucose infusion in mice: a new model to induce β-cell replication. Diabetes 2007; 56(7): 1792-801.
[http://dx.doi.org/10.2337/db06-1513] [PMID: 17400928]

[76]   Salpeter SJ, Klochendler A, Weinberg-Corem N, *et al.* Glucose regulates cyclin D2 expression in quiescent and replicating pancreatic β-cells through glycolysis and calcium channels. Endocrinology 2011; 152(7): 2589-98.
[http://dx.doi.org/10.1210/en.2010-1372] [PMID: 21521747]

[77]   Porat S, Weinberg-Corem N, Tornovsky-Babaey S, *et al.* Control of pancreatic β cell regeneration by glucose metabolism. Cell Metab 2011; 13(4): 440-9.
[http://dx.doi.org/10.1016/j.cmet.2011.02.012] [PMID: 21459328]

[78]   Novelli M, De Tata V, Bombara M, Bergamini E, Masiello P. Age-dependent reduction in GLUT-2 levels is correlated with the impairment of the insulin secretory response in isolated islets of Sprague-Dawley rats. Exp Gerontol 2000; 35(5): 641-51.
[http://dx.doi.org/10.1016/S0531-5565(00)00100-5] [PMID: 10978685]

[79]   Assefa Z, Lavens A, Steyaert C, *et al.* Glucose regulates rat β cell number through age-dependent effects on β cell survival and proliferation. PLoS One 2014; 9(1): e85174.
[http://dx.doi.org/10.1371/journal.pone.0085174] [PMID: 24416358]

[80]   Nikolova G, Jabs N, Konstantinova I, *et al.* The vascular basement membrane: a niche for insulin gene expression and β cell proliferation. Dev Cell 2006; 10(3): 397-405.
[http://dx.doi.org/10.1016/j.devcel.2006.01.015] [PMID: 16516842]

[81]   Nicholson JM, Arany EJ, Hill DJ. Changes in islet microvasculature following streptozotocin-induced β-cell loss and subsequent replacement in the neonatal rat. Exp Biol Med (Maywood) 2010; 235(2): 189-98.

[http://dx.doi.org/10.1258/ebm.2009.009316] [PMID: 20404034]

[82]   Almaça J, Molina J, Arrojo E Drigo R, *et al.* Young capillary vessels rejuvenate aged pancreatic islets. Proc Natl Acad Sci USA 2014; 111(49): 17612-7.
[http://dx.doi.org/10.1073/pnas.1414053111] [PMID: 25404292]

[83]   Rankin MM, Kushner JA. Aging induces a distinct gene expression program in mouse islets. Islets 2010; 2(6): 345-52.
[http://dx.doi.org/10.4161/isl.2.6.13376] [PMID: 21099336]

[84]   Krishnamurthy J, Torrice C, Ramsey MR, *et al.* Ink4a/Arf expression is a biomarker of aging. J Clin Invest 2004; 114(9): 1299-307.
[http://dx.doi.org/10.1172/JCI22475] [PMID: 15520862]

[85]   Dhawan S, Tschen SI, Bhushan A. Bmi-1 regulates the Ink4a/Arf locus to control pancreatic β-cell proliferation. Genes Dev 2009; 23(8): 906-11.
[http://dx.doi.org/10.1101/gad.1742609] [PMID: 19390085]

[86]   Cao R, Tsukada Y, Zhang Y. Role of Bmi-1 and Ring1A in H2A ubiquitylation and Hox gene silencing. Mol Cell 2005; 20(6): 845-54.
[http://dx.doi.org/10.1016/j.molcel.2005.12.002] [PMID: 16359901]

[87]   Cao R, Wang L, Wang H, *et al.* Role of histone H3 lysine 27 methylation in Polycomb-group silencing. Science 2002; 298(5595): 1039-43.
[http://dx.doi.org/10.1126/science.1076997] [PMID: 12351676]

[88]   Czermin B, Melfi R, McCabe D, Seitz V, Imhof A, Pirrotta V. Drosophila enhancer of Zeste/ESC complexes have a histone H3 methyltransferase activity that marks chromosomal Polycomb sites. Cell 2002; 111(2): 185-96.
[http://dx.doi.org/10.1016/S0092-8674(02)00975-3] [PMID: 12408863]

[89]   Kuzmichev A, Nishioka K, Erdjument-Bromage H, Tempst P, Reinberg D. Histone methyltransferase activity associated with a human multiprotein complex containing the Enhancer of Zeste protein. Genes Dev 2002; 16(22): 2893-905.
[http://dx.doi.org/10.1101/gad.1035902] [PMID: 12435631]

[90]   Müller J, Hart CM, Francis NJ, *et al.* Histone methyltransferase activity of a Drosophila Polycomb group repressor complex. Cell 2002; 111(2): 197-208.
[http://dx.doi.org/10.1016/S0092-8674(02)00976-5] [PMID: 12408864]

[91]   Hernández-Muñoz I, Taghavi P, Kuijl C, Neefjes J, van Lohuizen M. Association of BMI1 with polycomb bodies is dynamic and requires PRC2/EZH2 and the maintenance DNA methyltransferase DNMT1. Mol Cell Biol 2005; 25(24): 11047-58.
[http://dx.doi.org/10.1128/MCB.25.24.11047-11058.2005] [PMID: 16314526]

[92]   Chen H, Gu X, Su IH, *et al.* Polycomb protein Ezh2 regulates pancreatic β-cell Ink4a/Arf expression and regeneration in diabetes mellitus. Genes Dev 2009; 23(8): 975-85.
[http://dx.doi.org/10.1101/gad.1742509] [PMID: 19390090]

[93]   Zhou JX, Dhawan S, Fu H, *et al.* Combined modulation of polycomb and trithorax genes rejuvenates β cell replication. J Clin Invest 2013; 123(11): 4849-58.
[http://dx.doi.org/10.1172/JCI69468] [PMID: 24216481]

[94]   Drucker DJ, Buse JB, Taylor K, *et al.* Exenatide once weekly versus twice daily for the treatment of type 2 diabetes: a randomised, open-label, non-inferiority study. Lancet 2008; 372(9645): 1240-50.
[http://dx.doi.org/10.1016/S0140-6736(08)61206-4] [PMID: 18782641]

[95]   Bunck MC, Cornér A, Eliasson B, *et al.* Effects of exenatide on measures of β-cell function after 3 years in metformin-treated patients with type 2 diabetes. Diabetes Care 2011; 34(9): 2041-7.
[http://dx.doi.org/10.2337/dc11-0291] [PMID: 21868779]

[96]   Ghazi T, Rink L, Sherr JL, Herold KC. Acute metabolic effects of exenatide in patients with type 1 diabetes with and without residual insulin to oral and intravenous glucose challenges. Diabetes Care 2014; 37(1): 210-6.
[http://dx.doi.org/10.2337/dc13-1169] [PMID: 23939544]

[97]   Karatas H, Townsend EC, Cao F, *et al.* High-affinity, small-molecule peptidomimetic inhibitors of MLL1/WDR5 protein-protein interaction. J Am Chem Soc 2013; 135(2): 669-82.
[http://dx.doi.org/10.1021/ja306028q] [PMID: 23210835]

[98]   Keilhack H, Smith JJ. Small molecule inhibitors of EZH2: the emerging translational landscape. Epigenomics 2015; 7(3): 337-41.
[http://dx.doi.org/10.2217/epi.15.14] [PMID: 26077423]

CHAPTER 3

Human β-Cell Mass and Distribution in Health, Aging and Diabetes

Jonas L. Fowler, Ananta Poudel and **Manami Hara**[*]

Department of Medicine, The University of Chicago, Chicago, USA

Abstract: Regulation of pancreatic β-cell mass is an essential matter to understand pathophysiology of diabetes. Physiological and pathological changes of β-cell mass associated with aging, obesity and diabetes have been reported for over a century. However, the degree of compensation or alteration significantly varies among literature. The difficulty in studying the human pancreas is its large size and uneven distribution of β-cells/islets. Whole pancreas analysis has revealed intra-individual (regional) and inter-individual heterogeneity in β-cell mass, which hampers accurate quantification. Furthermore, physical β-cell loss is not the only contributing factor, but "dysfunctional" β-cells may be involved in insulin deficiency as well. Development of a practical stereological approach to quantify β-cell mass to overcome intra-individual and inter-individual heterogeneity would provide a standardized methodology in the field. Identification of marker(s) for quantifying dysfunctional β-cells that synthesize insulin but are deficient in insulin secretion should lead to a better understanding of β-cell pathophysiology.

Keywords: Aging, β-cell mass, Diabetes, Islets.

INTRODUCTION

Accurate quantification of β-cell mass in the human pancreas is challenging due to its large size. Furthermore, there is marked variability in the β-cell/islet distribution in the different regions of the pancreas. The head region is anatomically and developmentally distinct, and it exclusively contains a pancreatic polypeptide

---
[*] **Corresponding author Manami Hara:** Department of Medicine, The University of Chicago, Chicago, USA; Tel/Fax: 773-702-3727/773-834-0486; E-mail: mhara@uchicago.edu

David J. Hill (Ed.)

(PP) cell rich area, which can confound analysis of β-cell mass if not analyzed separately. Besides these regional differences, there is also marked variability among individuals that show no direct association with sex, age, BMI or type 2 diabetes (T2D). Starting with the basics of the human pancreas from its anatomy to intrinsic heterogeneity in β-cell mass, this chapter will review recent studies on β-cell mass and distribution in health, aging and diabetes and current difficulties of accurately assessing β-cell mass will be discussed.

HUMAN PANCREAS

Anatomy

The human pancreas is commonly divided into three main regions based on the anatomy, which are referred to the head, body, and tail. The head is localized closely to the duodenum on the right side of the abdomen. The superior mesenteric vessels beneath the neck are inferior and to the left of the head of the pancreas, while the uncinate process lies within the head posterior and to the right of the superior mesenteric vessels. The body region extends laterally to the left of the head, lying along the floor of the lesser sac, which is covered by peritoneum derived from the superior leaf of the transverse mesocolon. The tail region extends from the body laterally into the left side of the abdomen towards the spleen, where there is no coverage from transverse mesocolon derived peritoneum.

The head region is distinct from the rest of the pancreas both developmentally and anatomically [1] (Fig. **1**). The pancreas has two main sources of blood supply. The superior pancreaticoduodenal artery supplies the head region, while the splenic artery supplies both the body and tail. Hepatic ducts from the liver and cystic duct from the gall bladder merge into the common bile duct, which flows down to the head region of the pancreas and at the ampulla of Vater connects with the pancreatic duct. The ductal network is controlled by innervations from the celiac plexus and vagus. The sphincter of Oddi, a circular muscle band, functions as a valve to regulate the flow of pancreatic enzymes and bile into the duodenum, separating the pancreas from the intestinal environment.

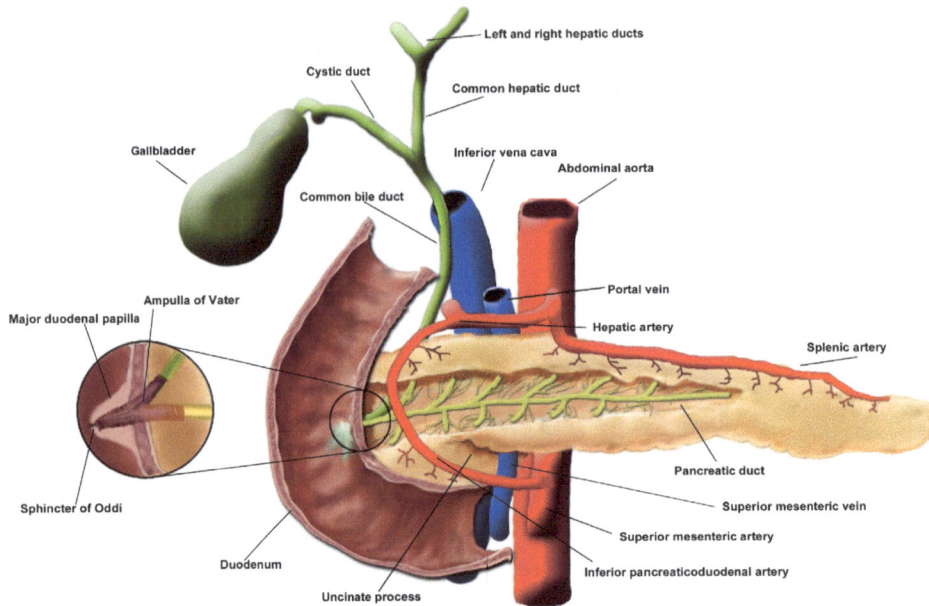

Fig. (1). Head region of the human pancreas (Reproduced from [1]).

Regional Differences in β-Cell/Islet Mass

The pancreatic islet is a highly vascularized micro-organ that is composed of multiple cell types including relatively large populations of β-, α- and δ-cells and small populations of PP- and ε-cells. There is an intrinsic regional difference in β-cell mass and correspondingly islet mass [2, 3]. While it is comparable between the head and body region, there is a gradual increase in β-cell and islet mass toward the tail region resulting in a ~2-fold difference. This regional difference in β-cell/islet density has important implications for the assessment of β-cell mass. Random inter-specimen comparisons between different regions may confound study accuracy. For example, a study of the tail region alone where the measured values are normalized using the pancreas volume or weight could result in overestimation of β-cell/islet mass, whereas using only the head or body regions could result in underestimation.

The Head of the Pancreas

The head of the pancreas differs from the rest of the pancreas in both anatomy and development. While the developmental origin of the body and tail region is the dorsal pancreatic bud, the head region is derived from the ventral bud. The head region exclusively contains a PP-cell rich area, which according to past studies covers 55-90% of the head region [4 - 8]. However, upon closer examination, we have found that the PP-cell rich area is largely restricted to the uncinate process [2]. Developmentally, the uncinate process originates from the ventral bud after fusion with the dorsal bud, following a ~270° anticlockwise rotation of the intestine. It is considered that an incomplete rotation results in improper growth and positioning of the dorsal or ventral bud, which leads to abnormal growth of the head with hypoplasia or aplasia of the uncinate process. Recent clinical studies on the diagnosis of pancreatic cancer have indicated the importance of understanding the physiology and development of the uncinate process. The dense expression of somatostatin receptors in PP-cells resulted in a common false-positive uptake of a tracer ^{68}Ga-DOTA-TOC by the uncinate process [9]. Hyperplasia of the head region, arisen from early development described above, can be misdiagnosed as pancreatic tumors [10].

The PP-cell distribution is low throughout the pancreas, except the PP-cell rich area where a distinct boundary exists between rich and poor regions [2]. In addition, the PP-cell rich area features a significant decrease in β- and α-cell mass, which can potentially affect β-cell mass quantification of the head region. Thus, it is critical that the PP-cell rich area is identified and assessed separately when quantifying β-cell mass in the head of the pancreas.

The head of the pancreas may be more prone to damage from the deterioration of surrounding tissues due to its anatomical proximity to the duodenum. In support of this notion, we have found that T2D patients undergo a preferential loss of β-cells in the head of the pancreas compared to the body and tail [3]. In addition, the majority of pancreatic tumors are of exocrine tissue with 95% of them being adenocarcinomas found in the head [11 - 15]. It is noted that the risk of developing pancreatic cancer increases 2-fold in individuals with diabetes [15, 16]. Endocrine tumors such as insulinoma only occur in 1 out of 250,000 individuals per year [17 - 21]. Furthermore, reports have indicated that there may be a link between the etiology of T1D and the dysfunction of the sphincter of Oddi in the head of the pancreas resulting in the infiltration of microbiota from the duodenum into the pancreatic duct. This infiltration causes non-specific inflammation in the exocrine and endocrine tissues resulting in an innate immune response that triggers the onset of T1D [22, 23]. Ultimately, the head of the pancreas has unique developmental and anatomical features that may make it more prone to various diseases.

Islet Architecture and Cellular Composition

There is a wide distribution of islet sizes ranging from single endocrine cells to larger islets consisting of several thousand cells, and this distribution is similar across different species. The conservation of the islet size suggests certain regulatory mechanisms to ensure proper function of islets [24]. Despite a similarity in size, the architecture of the islet does vary across species. Mouse islets are comprised of a β-cell core with a surrounding periphery of α- and δ-cells. This architecture can be seen in small human islets. However, larger human islets have a unique architecture that features a higher fraction of α- and δ-cells intermingled with β-cells [25]. This intermingled architecture can be observed in mice under insulin resistance such as pregnancy, obesity, diabetes and inflammation [24].

Mouse islets are mostly comprised of β-cells regardless of size. However, human islets exhibit a decrease in β-cell fraction coupled with an increase in α- and δ-cell fraction to go along with the unique intermingled architecture of larger islets. Collectively, human islet architecture and cellular composition are size-dependent. It is noted that smaller islets (including singlets and small clusters) are 9 times more frequent than larger islets. However, despite a higher prevalence of small islets, the large islets (>100 μm in diameter, less than 10% in number) generally comprise>40% of the islet mass [25]. There is no marked regional difference in this size-dependent distribution of endocrine cell composition. We have also shown that islets isolated separately from the head and body/tail regions function similarly in glucose-stimulated insulin secretion [3].

β-CELL MASS IN AGING AND DIABETES

Absolute Mass of β-Cells and Islets in Humans

It is of great interest to determine absolute mass of β-cells/islets in the human pancreas. Since pancreas weight is generally considered to correlate with individual body weight, marked variability is anticipated. Early studies reported the pancreas weight in the range of 30-150 g [26]. Rahier *et al.* reported the similar range of pancreas weight of 82 g average (n=8, 67-110 g) for control subjects, the range of 26-51 g for patients with T1D (n=4), and 55-100 g for patients with T2D (n=8) [27]. They further calculated an average total mass of endocrine tissue based on the volume density (using the Chalkley point counting methodology [28]): controls 1,395 mg, T2D 1,449 mg and T1D 413 mg, with β-cell mass about 800 mg both in controls and T2D patients.

β-Cell Mass in Aging and Obesity

It is still unclear how normal aging affects β-cell mass. Aging-related pancreas atrophy, particularly in the elderly may lead to a gradual decline in β-cell mass [29, 30]. Rahier *et al.* reported that there was a small age-dependent decrease in β-cell mass between two non-diabetic subject groups (40-60 years and 70-90 years; n=52) [29]. Saisho *et al.* recently showed that β-cell mass remained constant from age 20-102 despite exocrine pancreas atrophy with the only notable change being an age-dependent increase in β-cell nuclear diameter [30].

Several studies found a moderate to marked correlation between β-cell mass and body mass index (BMI) [30 - 33]. Rahier *et al.* reported a 20% increase in β-cell mass in obese individuals of European origin [29]. Saisho *et al.* reported a 50% increase in β-cell mass in obese *versus* non-obese individuals [30]. However, a study in Japanese lean (n=39) and obese (n=33) individuals found no difference in β-cell mass or β-cell turnover [33].

Loss of β-Cell Mass in Diabetes

Type 1 diabetes (T1D) has a significant impact on β-cell mass with a nearly complete loss of β-cells resulting in an absolute insulin deficiency [27, 34 - 38], consistent with studies on insulin secretion [39 - 42]. The emerging important notion is that there are fundamental differences in T1D between children and adults in regard to pathogenesis and disease progression. Early childhood-onset T1D patients exhibit severe β-cell depletion, whereas there is a greater population of residual β-cells in individuals with adult-onset T1D. A number of studies have demonstrated that, in addition to the presence of multiple antibodies and human leukocyte antigen (HLA) risk alleles, age of onset is an important risk factor in the severity of T1D progression [43 - 48]. Pipeleers *et al.* found that C-peptide secretion was markedly lower in T1D patients with onset under the age of 7-yr, indicating that β-cell depletion is more severe in earlier onset T1D [49]. It is noted that adult-onset autoimmune diabetes is complex and often diagnosed as T2D. We carried out whole pancreas analyses of T1D with childhood and adult onset cases [50]. In both cases, there were significant regional differences in β-cell loss. The most severe loss of β-cells occurred in the tail followed by the body and head regions. In addition, the islets containing β-cells were exclusively found in the head region of the adult-onset pancreas, while the rest of the adult pancreas as well as the entire childhood onset pancreas were completely lacking intra-islet β-cells.

Contribution of β-cell loss to T2D has also been demonstrated from moderate [29, 31, 51, 52] to marked (>40%) [32, 53 - 56], while some reported no significant differences compared to non-diabetic controls [27, 57]. We have reported ~45% decrease in β-cell mass in T2D patients compared to non-diabetic subjects, which is largely due to the loss of large islets resulting in a significant shift in the islet size distribution [58]. In the following study with more detailed regional analysis, we showed that the head of the pancreas exhibited a preferential loss of β-cells, while the body and tail regions had little change (T2D patients: n=12; non-diabetic subjects: n=23) [3].

Physical *vs.* Functional Loss of β-Cell Mass

While it is evident that a physical loss of insulin-producing β-cells occurs in patients with T2D, reduced β-cell mass does not solely explain the development of chronic hyperglycemia in T2D. In fact, it has been clinically well documented that hemipancreatectomy in healthy donors does not result in diabetes [59 - 64]. It may be reasonable to assume that "dysfunctional" β-cells that synthesize insulin but are deficient in secretion are spared for a relatively long period of time. This should explain clinical deteriorations of increased insulin resistance and reduced β-cell function associated with T2D, obesity and aging, which may not be directly reflected in a physical β-cell loss.

WHOLE PANCREAS ANALYSIS

Whole pancreas analysis has revealed marked intra-individual variability (*i.e.* regional differences), as well as inter-individual heterogeneity in β-cell/islet mass. In Fig. (**2**), we show ten cases of the whole pancreas analysis from head, body to tail region (from left to right in the X-axis); non-diabetic subjects in the left column (**A-E**) and patients with T2D in the right column (**F-J**) from the young to the old (from top to bottom) [65]. Pancreata were collected from organ donors for clinical transplantation with consent for research use. Each specimen was cut into consecutive 5 mm tissue blocks where the odd blocks were fresh-frozen and the even blocks were embedded in paraffin. Tissue sections (~ 5 μm in thickness) were immunostained for insulin, glucagon and somatostatin and labeled with multi-fluorescent secondary dyes.

An Olympus IX8 DSU spinning disk confocal microscope (Melville, NY) was used in conjunction with StereoInvestigator imaging software (SI, MicroBrightField, Williston, VT). The SI imaging software allows for 3-dimensional control over a motorized stage, resulting in the acquisition of consecutive images to be compiled into a high-resolution montage that represents multiple microscopic fields of view. We employed a modified method of "virtual slice capture" using a 10x objective in order to capture microscopic images [66, 67]. Virtual slices were obtained for the three different immunostained hormones as well as nucleus staining by DAPI in order to generate a single multichannel composite image. A custom-written macro for Fiji/ImageJ (http://rsbweb.nih.gov/ij/) was used for the analysis and quantification of cellular composition within the tissue.

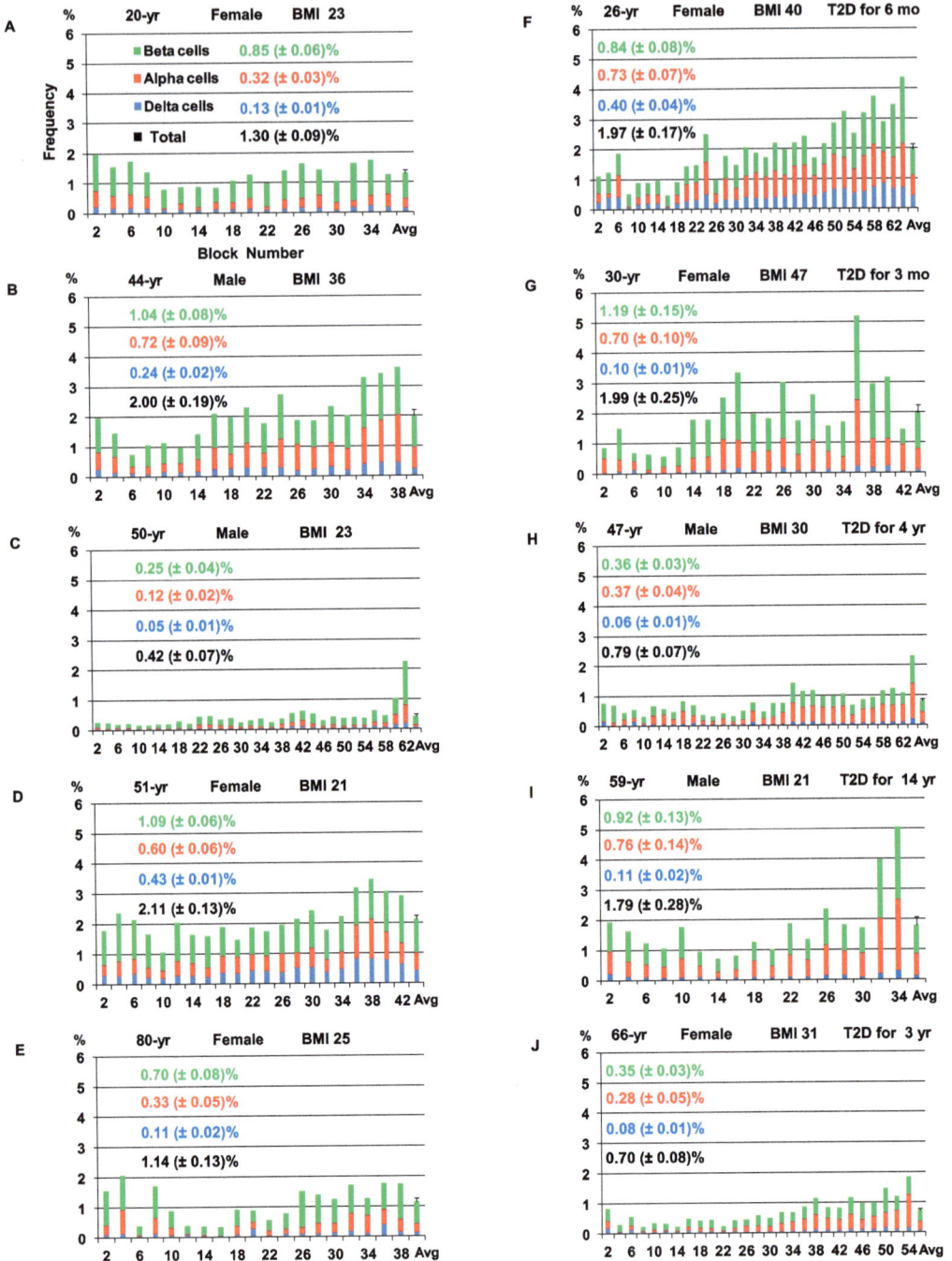

Fig. (2). Whole pancreas analysis. A-E: non-diabetic subjects. **F-J:** patients with T2D. X-axis: Block number from head, body to tail region (from left to right) (Reproduced from [65]).

All the graphs are plotted in the same scale for a better comparison. Overall, regional distribution of β-cell and islet mass shows a gradual increase from head to tail region, although it fluctuates markedly. Cases A and E contain similar mass between the head and tail region. There appears no difference regarding age and sex. Obese individuals, non-diabetic B and most of the patients with T2D except I, show no distinction from those with normal weight. Heterogeneity among patients with T2D, and further comparing to non-diabetic subjects, is striking. Such variability may suggest that each individual has a distinct baseline of β-cell and islet mass, which may have been determined in their early life [68]. Collectively, it is noted that sampling bias can occur in two dimensions stemming from: the intra-individual regional variability and the inter-individual heterogeneity.

CONCLUDING REMARKS

The difficulty of interpreting reported data and deducing a general consensus of β-cell mass in health, aging and diabetes largely lies in the intra-individual variability and the inter-individual heterogeneity, the latter is far more complex. It is thus noted that many studies may have been suffered from sampling bias, inaccurate quantification and improper comparison. Most importantly, it is critical to find parameter(s) to accurately assess individual β-cell and islet mass and identify proper control subjects in comparative studies to overcome the inter-individual heterogeneity. Furthermore, identification of markers for dysfunctional β-cells should advance our understanding of β-cell physiology and biology toward a cure for diabetes.

CONFLICT OF INTEREST

The authors confirm that they have no conflict of interest to declare for this publication.

ACKNOWLEDGEMENTS

Declared none.

REFERENCES

[1]   Savari O, Zielinski MC, Wang X, *et al*. Distinct function of the head region of human pancreas in the pathogenesis of diabetes. Islets 2013; 5(5): 226-8.

[http://dx.doi.org/10.4161/isl.26432] [PMID: 24045229]

[2]   Wang X, Zielinski MC, Misawa R, *et al.* Quantitative analysis of pancreatic polypeptide cell distribution in the human pancreas. PLoS One 2013; 8(1): e55501.
[http://dx.doi.org/10.1371/journal.pone.0055501] [PMID: 23383206]

[3]   Wang X, Misawa R, Zielinski MC, Shen J, Witkowski P, Hara M. Regional differences in islet size distribution and architecture in human adult pancreas. PLoS One 2013; 8: e67454.
[http://dx.doi.org/10.1371/journal.pone.0067454] [PMID: 23826303]

[4]   Orci L, Malaisse-Lagae F, Baetens D, Perrelet A. Pancreatic-polypeptide-rich regions in human pancreas. Lancet 1978; 2(1): 200-1.
[http://dx.doi.org/10.1016/S0140-6736(78)92181-5]

[5]   Malaisse-Lagae F, Stefan Y, Cox J, Perrelet A, Orci L. Identification of a lobe in the adult human pancreas rich in pancreatic polypeptide. Diabetologia 1979; 17(6): 361-5.
[http://dx.doi.org/10.1007/BF01236270] [PMID: 395002]

[6]   Gersell DJ, Gingerich RL, Greider MH. Regional distribution and concentration of pancreatic polypeptide in the human and canine pancreas. Diabetes 1979; 28(1): 11-5.
[http://dx.doi.org/10.2337/diabetes.28.1.11] [PMID: 759245]

[7]   Stefan Y, Orci L, Malaisse-Lagae F, Perrelet A, Patel Y, Unger RH. Quantitation of endocrine cell content in the pancreas of nondiabetic and diabetic humans. Diabetes 1982; 31(8 Pt 1): 694-700.
[http://dx.doi.org/10.2337/diab.31.8.694] [PMID: 6131002]

[8]   Rahier J, Wallon J, Loozen S, Lefevre A, Gepts W, Haot J. The pancreatic polypeptide cells in the human pancreas: the effects of age and diabetes. J Clin Endocrinol Metab 1983; 56(3): 441-4.
[http://dx.doi.org/10.1210/jcem-56-3-441] [PMID: 6337179]

[9]   Jacobsson H, Larsson P, Jonsson C, Jussing E, Grybäck P. Normal uptake of 68Ga-DOTA-TOC by the pancreas uncinate process mimicking malignancy at somatostatin receptor PET. Clin Nucl Med 2012; 37(4): 362-5.
[http://dx.doi.org/10.1097/RLU.0b013e3182485110] [PMID: 22391705]

[10]   Chandra J, Grierson C, Bungay H. Normal variations in pancreatic contour are associated with intestinal malrotation and can mimic neoplasm. Clin Radiol 2012; 67(12): 1187-92.
[http://dx.doi.org/10.1016/j.crad.2011.11.021] [PMID: 22766483]

[11]   Nathan H, Wolfgang CL, Edil BH, *et al.* Peri-operative mortality and long-term survival after total pancreatectomy for pancreatic adenocarcinoma: a population-based perspective. J Surg Oncol 2009; 99(2): 87-92.
[http://dx.doi.org/10.1002/jso.21189] [PMID: 19021191]

[12]   Sata N, Kurashina K, Nagai H, *et al.* The effect of adjuvant and neoadjuvant chemo(radio)therapy on survival in 1,679 resected pancreatic carcinoma cases in Japan: report of the national survey in the 34th annual meeting of Japanese Society of Pancreatic Surgery. J Hepatobiliary Pancreat Surg 2009; 16(4): 485-92.
[http://dx.doi.org/10.1007/s00534-009-0077-7] [PMID: 19333537]

[13]   Lau MK, Davila JA, Shaib YH. Incidence and survival of pancreatic head and body and tail cancers: a population-based study in the United States. Pancreas 2010; 39(4): 458-62.

[http://dx.doi.org/10.1097/MPA.0b013e3181bd6489] [PMID: 19924019]

[14]　Bouvier AM, David M, Jooste V, Chauvenet M, Lepage C, Faivre J. Rising incidence of pancreatic cancer in France. Pancreas 2010; 39(8): 1243-6.
[http://dx.doi.org/10.1097/MPA.0b013e3181e1d5b3] [PMID: 20881902]

[15]　Bartosch-Härlid A, Andersson R. Diabetes mellitus in pancreatic cancer and the need for diagnosis of asymptomatic disease. Pancreatology 2010; 10(4): 423-8.
[http://dx.doi.org/10.1159/000264676] [PMID: 20720443]

[16]　Cui Y, Andersen DK. Diabetes and pancreatic cancer. Endocr Relat Cancer 2012; 19(5): F9-F26.
[http://dx.doi.org/10.1530/ERC-12-0105] [PMID: 22843556]

[17]　Placzkowski KA, Vella A, Thompson GB, *et al.* Secular trends in the presentation and management of functioning insulinoma at the Mayo Clinic, 1987-2007. J Clin Endocrinol Metab 2009; 94(4): 1069-73.
[http://dx.doi.org/10.1210/jc.2008-2031] [PMID: 19141587]

[18]　Service FJ, McMahon MM, OBrien PC, Ballard DJ. Functioning insulinomaincidence, recurrence, and long-term survival of patients: a 60-year study. Mayo Clin Proc 1991; 66(7): 711-9.
[http://dx.doi.org/10.1016/S0025-6196(12)62083-7] [PMID: 1677058]

[19]　Service FJ, Dale AJ, Elveback LR, Jiang NS. Insulinoma: clinical and diagnostic features of 60 consecutive cases. Mayo Clin Proc 1976; 51(7): 417-29.
[PMID: 180358]

[20]　Kavlie H, White TT. Non-beta hormone-producing islet cell tumors of the pancreas. Am Surg 1972; 38(11): 601-7.
[PMID: 4343133]

[21]　Cullen RM, Ong CE. Insulinoma in Auckland 1970-1985. N Z Med J 1987; 100(831): 560-2.
[PMID: 2836767]

[22]　Korsgren S, Molin Y, Salmela K, Lundgren T, Melhus A, Korsgren O. On the etiology of type 1 diabetes: a new animal model signifying a decisive role for bacteria eliciting an adverse innate immunity response. Am J Pathol 2012; 181(5): 1735-48.
[http://dx.doi.org/10.1016/j.ajpath.2012.07.022] [PMID: 22944599]

[23]　Skog O, Korsgren S, Melhus A, Korsgren O. Revisiting the notion of type 1 diabetes being a T-cel--mediated autoimmune disease. Curr Opin Endocrinol Diabetes Obes 2013; 20(2): 118-23.
[http://dx.doi.org/10.1097/MED.0b013e32835edb89] [PMID: 23422243]

[24]　Kim A, Miller K, Jo J, Kilimnik G, Wojcik P, Hara M. Islet architecture: A comparative study. Islets 2009; 1(2): 129-36.
[http://dx.doi.org/10.4161/isl.1.2.9480] [PMID: 20606719]

[25]　Kilimnik G, Jo J, Periwal V, Zielinski MC, Hara M. Quantification of islet size and architecture. Islets 2012; 4(2): 167-72.
[http://dx.doi.org/10.4161/isl.19256] [PMID: 22653677]

[26]　Schaefer JH. The normal weight of the pancreas in the adult human being: A biometric study. Anat Rec 1926; 32: 119-32.
[http://dx.doi.org/10.1002/ar.1090320204]

[27]   Rahier J, Goebbels RM, Henquin JC. Cellular composition of the human diabetic pancreas. Diabetologia 1983; 24(5): 366-71.
[http://dx.doi.org/10.1007/BF00251826] [PMID: 6347784]

[28]   Chalkley HW. Method for the quantitative morphologic analysis of tissues. Natl Cancer Inst 1943; 4: 47-53.

[29]   Rahier J, Guiot Y, Goebbels RM, Sempoux C, Henquin JC. Pancreatic β-cell mass in European subjects with type 2 diabetes. Diabetes Obes Metab 2008; 10 (Suppl. 4): 32-42.
[http://dx.doi.org/10.1111/j.1463-1326.2008.00969.x] [PMID: 18834431]

[30]   Saisho Y, Butler AE, Manesso E, Elashoff D, Rizza RA, Butler PC. β-cell mass and turnover in humans: effects of obesity and aging. Diabetes Care 2013; 36(1): 111-7.
[http://dx.doi.org/10.2337/dc12-0421] [PMID: 22875233]

[31]   Yoon KH, Ko SH, Cho JH, *et al.* Selective β-cell loss and alpha-cell expansion in patients with type 2 diabetes mellitus in Korea. J Clin Endocrinol Metab 2003; 88(5): 2300-8.
[http://dx.doi.org/10.1210/jc.2002-020735] [PMID: 12727989]

[32]   Butler AE, Janson J, Bonner-Weir S, Ritzel R, Rizza RA, Butler PC. β-cell deficit and increased β-cell apoptosis in humans with type 2 diabetes. Diabetes 2003; 52(1): 102-10.
[http://dx.doi.org/10.2337/diabetes.52.1.102] [PMID: 12502499]

[33]   Kou K, Saisho Y, Satoh S, Yamada T, Itoh H. Change in β-cell mass in Japanese nondiabetic obese individuals. J Clin Endocrinol Metab 2013; 98(9): 3724-30.
[http://dx.doi.org/10.1210/jc.2013-1373] [PMID: 23766518]

[34]   MacLean N, Ogilvie RF. Observations on the pancreatic islet tissue of young diabetic subjects. Diabetes 1959; 8(2): 83-91.
[http://dx.doi.org/10.2337/diab.8.2.83] [PMID: 13630179]

[35]   Gepts W. Pathologic anatomy of the pancreas in juvenile diabetes mellitus. Diabetes 1965; 14(10): 619-33.
[http://dx.doi.org/10.2337/diab.14.10.619] [PMID: 5318831]

[36]   Doniach I, Morgan AG. Islets of Langerhans in juvenile diabetes mellitus. Clin Endocrinol (Oxf) 1973; 2(3): 233-48.
[http://dx.doi.org/10.1111/j.1365-2265.1973.tb00425.x] [PMID: 4586527]

[37]   Madsbad S, Faber OK, Binder C, McNair P, Christiansen C, Transbøl I. Prevalence of residual β-cell function in insulin-dependent diabetics in relation to age at onset and duration of diabetes. Diabetes 1978; 27 (Suppl. 1): 262-4.
[http://dx.doi.org/10.2337/diab.27.1.S262] [PMID: 344117]

[38]   Oram RA, Jones AG, Besser RE, *et al.* The majority of patients with long-duration type 1 diabetes are insulin microsecretors and have functioning beta cells. Diabetologia 2014; 57(1): 187-91.
[http://dx.doi.org/10.1007/s00125-013-3067-x] [PMID: 24121625]

[39]   Levetan C. Distinctions between islet neogenesis and β-cell replication: implications for reversal of Type 1 and 2 diabetes. J Diabetes 2010; 2(2): 76-84.
[http://dx.doi.org/10.1111/j.1753-0407.2010.00074.x] [PMID: 20923488]

[40] Keenan HA, Sun JK, Levine J, *et al.* Residual insulin production and pancreatic ß-cell turnover after 50 years of diabetes: Joslin medalist study. Diabetes 2010; 59(11): 2846-53.
[http://dx.doi.org/10.2337/db10-0676] [PMID: 20699420]

[41] Löhr M, Klöppel G. Residual insulin positivity and pancreatic atrophy in relation to duration of chronic type 1 (insulin-dependent) diabetes mellitus and microangiopathy. Diabetologia 1987; 30(10): 757-62.
[http://dx.doi.org/10.1007/BF00275740] [PMID: 3322901]

[42] Matveyenko AV, Butler PC. Relationship between β-cell mass and diabetes onset. Diabetes Obes Metab 2008; 10 (Suppl. 4): 23-31.
[http://dx.doi.org/10.1111/j.1463-1326.2008.00939.x] [PMID: 18834430]

[43] Karjalainen J, Salmela P, Ilonen J, Surcel HM, Knip M. A comparison of childhood and adult type I diabetes mellitus. N Engl J Med 1989; 320(14): 881-6.
[http://dx.doi.org/10.1056/NEJM198904063201401] [PMID: 2648146]

[44] Schiffrin A, Suissa S, Weitzner G, Poussier P, Lalla D. Factors predicting course of β-cell function in IDDM. Diabetes Care 1992; 15(8): 997-1001.
[http://dx.doi.org/10.2337/diacare.15.8.997] [PMID: 1505333]

[45] Sabbah E, Savola K, Ebeling T, *et al.* Genetic, autoimmune, and clinical characteristics of childhood- and adult-onset type 1 diabetes. Diabetes Care 2000; 23(9): 1326-32.
[http://dx.doi.org/10.2337/diacare.23.9.1326] [PMID: 10977027]

[46] Tsai EB, Sherry NA, Palmer JP, Herold KC. The rise and fall of insulin secretion in type 1 diabetes mellitus. Diabetologia 2006; 49(2): 261-70.
[http://dx.doi.org/10.1007/s00125-005-0100-8] [PMID: 16404554]

[47] Gallagher MP, Goland RS, Greenbaum CJ. Making progress: preserving beta cells in type 1 diabetes. Ann N Y Acad Sci 2011; 1243: 119-34.
[http://dx.doi.org/10.1111/j.1749-6632.2011.06321.x] [PMID: 22211897]

[48] Merger SR, Leslie RD, Boehm BO. The broad clinical phenotype of Type 1 diabetes at presentation. Diabet Med 2013; 30(2): 170-8.
[http://dx.doi.org/10.1111/dme.12048] [PMID: 23075321]

[49] Pipeleers D, Chintinne M, Denys B, Martens G, Keymeulen B, Gorus F. Restoring a functional β-cell mass in diabetes. Diabetes Obes Metab 2008; 10(10) (Suppl. 4): 54-62.
[http://dx.doi.org/10.1111/j.1463-1326.2008.00941.x] [PMID: 18834433]

[50] Poudel A, Savari O, Striegel DA, *et al.* β-cell destruction and preservation in childhood and adult onset type 1 diabetes. Endocrine 2015; 49(3): 693-702.
[http://dx.doi.org/10.1007/s12020-015-0534-9]

[51] Clark A, Wells CA, Buley ID, *et al.* Islet amyloid, increased A-cells, reduced B-cells and exocrine fibrosis: quantitative changes in the pancreas in type 2 diabetes. Diabetes Res 1988; 9(4): 151-9.
[PMID: 3073901]

[52] Sakuraba H, Mizukami H, Yagihashi N, Wada R, Hanyu C, Yagihashi S. Reduced β-cell mass and expression of oxidative stress-related DNA damage in the islet of Japanese Type II diabetic patients. Diabetologia 2002; 45(1): 85-96.

[http://dx.doi.org/10.1007/s125-002-8248-z] [PMID: 11845227]

[53]   MacLean N, Ogilvie RF. Quantitative estimation of the pancreatic islet tissue in diabetic subjects. Diabetes 1955; 4(5): 367-76.
[http://dx.doi.org/10.2337/diab.4.5.367] [PMID: 13270659]

[54]   Saito K, Yaginuma N, Takahashi T. Differential volumetry of A, B and D cells in the pancreatic islets of diabetic and nondiabetic subjects. Tohoku J Exp Med 1979; 129(3): 273-83.
[http://dx.doi.org/10.1620/tjem.129.273] [PMID: 392812]

[55]   Stefan Y, Orci L, Malaisse-Lagae F, Perrelet A, Patel Y, Unger RH. Quantitation of endocrine cell content in the pancreas of nondiabetic and diabetic humans. Diabetes 1982; 31(8 Pt 1): 694-700.
[http://dx.doi.org/10.2337/diab.31.8.694] [PMID: 6131002]

[56]   Klöppel G, Löhr M, Habich K, Oberholzer M, Heitz PU. Islet pathology and the pathogenesis of type 1 and type 2 diabetes mellitus revisited. Surv Synth Pathol Res 1985; 4(2): 110-25.
[PMID: 3901180]

[57]   Clark A, Jones LC, de Koning E, Hansen BC, Matthews DR. Decreased insulin secretion in type 2 diabetes: a problem of cellular mass or function? Diabetes 2001; 50 (Suppl. 1): S169-71.
[http://dx.doi.org/10.2337/diabetes.50.2007.S169] [PMID: 11272183]

[58]   Kilimnik G, Zhao B, Jo J, et al. Altered islet composition and disproportionate loss of large islets in patients with type 2 diabetes. PLoS One 2011; 6(11): e27445.
[http://dx.doi.org/10.1371/journal.pone.0027445] [PMID: 22102895]

[59]   Kendall DM, Sutherland DE, Goetz FC, Najarian JS. Metabolic effect of hemipancreatectomy in donors. Preoperative prediction of postoperative oral glucose tolerance. Diabetes 1989; 38 (Suppl. 1): 101-3.
[http://dx.doi.org/10.2337/diab.38.1.S101] [PMID: 2491996]

[60]   Kendall DM, Sutherland DE, Najarian JS, Goetz FC, Robertson RP. Effects of hemipancreatectomy on insulin secretion and glucose tolerance in healthy humans. N Engl J Med 1990; 322(13): 898-903.
[http://dx.doi.org/10.1056/NEJM199003293221305] [PMID: 2179721]

[61]   Seaquist ER, Robertson RP. Effects of hemipancreatectomy on pancreatic alpha and beta cell function in healthy human donors. J Clin Invest 1992; 89(6): 1761-6.
[http://dx.doi.org/10.1172/JCI115779] [PMID: 1601986]

[62]   Robertson RP, Lanz KJ, Sutherland DE, Seaquist ER. Relationship between diabetes and obesity 9 to 18 years after hemipancreatectomy and transplantation in donors and recipients. Transplantation 2002; 73(5): 736-41.
[http://dx.doi.org/10.1097/00007890-200203150-00013] [PMID: 11907419]

[63]   Reynoso JF, Gruessner CE, Sutherland DE, Gruessner RW. Short- and long-term outcome for living pancreas donors. J Hepatobiliary Pancreat Sci 2010; 17(2): 92-6.
[http://dx.doi.org/10.1007/s00534-009-0147-x] [PMID: 19652901]

[64]   Sutherland DE, Radosevich D, Gruessner R, Gruessner A, Kandaswamy R. Pushing the envelope: living donor pancreas transplantation. Curr Opin Organ Transplant 2012; 17(1): 106-15.
[http://dx.doi.org/10.1097/MOT.0b013e32834ee6e5] [PMID: 22240639]

[65]   Sutherland DE, Poudel A, Fowler JL, et al. Steleological analyses of the whole human pancreas. Sci

Rep 2016; 6: 340-49.
[http://dx.doi.org/10.1038/srep34049] [PMID: 27658965]

[66]   Kilimnik G, Kim A, Jo J, Miller K, Hara M. Quantification of pancreatic islet distribution *in situ* in mice. Am J Physiol Endocrinol Metab 2009; 297(6): E1331-8.
[http://dx.doi.org/10.1152/ajpendo.00479.2009] [PMID: 19808908]

[67]   Miller K, Kim A, Kilimnik G, *et al.* Islet formation during the neonatal development in mice. PLoS One 2009; 4(11): e7739.
[http://dx.doi.org/10.1371/journal.pone.0007739] [PMID: 19893748]

[68]   Gregg BE, Moore PC, Demozay D, *et al.* Formation of a human β-cell population within pancreatic islets is set early in life. J Clin Endocrinol Metab 2012; 97(9): 3197-206.
[http://dx.doi.org/10.1210/jc.2012-1206] [PMID: 22745242]

CHAPTER 4

Gestational Programming of β-Cell Mass and Pancreatic Function in the Next Generation

David J. Hill*

Lawson Health Research Institute, St. Joseph's Health Care, 268 Grosvenor Street, London, Ontario N6A 4V2, Canada

Abstract: The gestational environment can have profound effects on the future health of the offspring, including a greater risk of type 2 diabetes and of cardiovascular diseases. Whilst the function of numerous tissues that can impact on future metabolism are altered by an adverse fetal environment, including the hypothalamic control of appetite and the release of glucocorticoids, hepatic function, and the insulin sensitive tissues such as skeletal muscle and adipose, some of the most definitive data concerns changes in the phenotype and function of the pancreatic β-cells. A number of animal models of intrauterine growth restriction (IUGR) have been utilized to study the long-term effects on the offspring, such as a reduced maternal calorie intake, a reduced protein content of the diet, uterine vessel occlusion, and nicotine administration. Changes to the pancreatic β-cells are remarkably similar and include a reduced tissue mass, lower rate of proliferation, increased developmental apoptosis, less plasticity following damage postnatally, higher sensitivity to cytotoxic cytokines, and reduced glucose-stimulated insulin release. These changes persist into adulthood and result in impaired glucose tolerance, Similar changes are also seen in offspring from pregnancies complicated by maternal diabetes. The mechanisms responsible for the altered β-cells function include changes to the mTOR signaling pathway, epigenetic changes altering the expression of key genes involved with β-cell growth and insulin synthesis, and changes in the rate of telomere shortening resulting in premature cellular aging. These pathways may also be influenced by environmental toxins during pregnancy. Nutritional intervention by micronutrient supplementation of the mother, or treatment of the newborn with peptide hormones trophic for the β-cells can reverse the pancreatic phenotype and reduce the risk of adult metabolic disease.

* **Corresponding author David J. Hill:** Lawson Health Research Institute, St. Joseph's Health Care, 268 Grosvenor Street, London, Ontario N6A 4V2, Canada; Tel/Fax: 519 6466100 Ext. 64716; E-mail: david.hill@lhrionhealth.ca

Keywords: β-cell mass, Epigenetics, Fetal programming, Gestational diabetes, Intra-uterine growth restriction, MTOR, Pregnancy, Type 2 diabetes.

INTRODUCTION

The pre-conceptual and gestational environment can have profound effects of the future health of the offspring, including a greater risk for type 2 diabetes, cardiovascular diseases and other chronic diseases [1]. This was convincingly shown in cohort studies of individuals exposed to the Dutch hunger winter of 1944/45 where severe maternal calorie restriction during first trimester resulted in a reduced birth weight and a higher rate of offspring obesity by 19 years of age, associated with diabetes and vascular disorders in later life [2]. Surprisingly, similar disease risks are observed in the offspring of women who were obese or had diabetes prior to conception, or who developed hyperglycemia, with or without gestational diabetes [3 - 6]. The impact of relative under- or over-nutrition of the fetus on future health has been shown to involve programmed changes to a number of key tissues involved with energy homeostasis, including adipose, liver, and muscle [7], as well as changes to the neuronal architecture of the hypothalamus resulting in altered production of peptides that determine appetite, such as NPY [8], responsiveness to leptin [9] and the basal levels of cortisol and tissue glucocorticoid receptors [10]. Fetal programming of future metabolic diseases is therefore likely to represent an accumulated effect of developmental changes across multiple tissues. However, some of the most profound and well-studied determinants involve changes to the phenotype and function of the pancreatic β-cells. This chapter will review some of the animal models used and the mechanisms proposed for environmental programming of the developing endocrine pancreas prior to birth, and some of the corrective strategies so far investigated.

ANIMAL MODELS OF FETAL PROGRAMMING OF THE ENDOCRINE PANCREAS THROUGH NUTRITIONAL DEFICIT

A number of animal models that disrupt maternal nutritional availability or uteroplacental function have been used to represent the human small-fo--gestational age term infant, as reviewed previously [11]. These include reduced calorie availability to the pregnant rat or mouse [12, 13], uterine vessel occlusion to reduce utero-placental blood flow [14] and nicotine administration to the mother [15]. In each case the resulting changes to endocrine pancreas morphology and function are remarkably similar. We and others have utilized a maternal low protein diet (LP) made isocalorific by the addition of carbohydrate from gestation to either parturition, or continued until weaning. If LP diet was given to rats or mice throughout pregnancy it resulted in a lower β-cell mass and mean islet size in the offspring at birth, and this was further exacerbated at weaning [16 - 19]. This was due to less β-cell proliferation together with an increased rate of developmental apoptosis. Further analysis showed that the cell cycle kinetics of β-cell replication had been altered with an extended G1 phase [16]. Whilst some recovery of β-cell mass was possible if the LP dietary insult was removed at parturition, severe deficits remained throughout life if the insult was extended to weaning [17]. Glucose-stimulated insulin release was reduced and β-cells were more susceptible to cytokine-induced cell death *in vitro* [16, 20]. These deficits were transmitted to the F2 generation through females, even when the F1 offspring received a normal diet through gestation. Once adult at 130 days of age the offspring of the LP-fed rats were glucose intolerant with peripheral glucose resistance [19]. A direct human correlate of the maternal LP diet model in rodents is the reduced birth weight associated with restricted maternal protein intake experienced by vegetarian women in rural India, as characterized by the Pune Maternal Nutrition Study [21, 22].

Exposure to LP diet during gestation also had phenotypic effects on the anatomy of the microvasculature of the pancreas. This is important as paracrine signaling occurs across the basement membrane juxtaposing the β-cell and the capillary endothelium, mediated by integrins and by various peptide growth factors, including vascular endothelial growth factor (VEGF) and hepatocyte growth factor (HGF) and [23 - 25]. The presence of VEGF-A is necessary to maintain a fenestrated islet endothelium that allows for rapid glucose sensing from the circulation and the export of secreted insulin [24]. We reported that intra-islet vascular volume and the abundance of the VEGF receptor were both lower in the offspring of LP-fed animals at birth [26]. Islet microvacular density remained compromised in the pancreas of the offspring until adulthood. Similarly, the vasculature of the islets was also less abundant in additional animal models of IUGR, including ligation of the uterine arteries [27]. The offspring of rats receiving LP diet during pregnancy also showed a reduced capillary density in a variety of non-pancreatic tissues, such as endometrium, ovaries and skeletal muscle [28]. Dilatation of the endothelial was impaired also [29]. We found that the number of endothelial progenitor cells (EPC) present in the pancreas, identified from the presence of nestin and CD34, was significantly lower in the offspring of animals receiving a LP diet [30]. This was accompanied by a reduction in the expression of IGF-II both in pancreas and some other tissues.

Several laboratories, including our own, have examined the impact of a gestational LP diet in the mouse on the plasticity of β-cells in the offspring to recover from experimental depletion using streptozotocin [STZ] [31, 32]. The ability of β-cells to near-fully regenerate mass following sub-total destruction normally seen in juveniles was abolished if the offspring had a prior exposure to LP diet [31]. Whether this is due primarily to a lack of plasticity in the β-cells or in the supporting microvasculature is not clear. However, paradoxically, there was an increase in the number of Pdx1$^+$/insulin$^-$ putative β-cell progenitors [31] in offspring of LP-fed mice following STZ, suggesting a possible inhibition on their differentiation into mature β-cells. Within 72h of STZ administration to offspring of control diet-fed mice a significant increase was seen in the number of insulin$^+$ cell clusters, or individual insulin$^+$ cells adjacent to the pancreatic ducts. This failed to occur in offspring of LP-fed animals, suggesting that previous protein insufficiency also compromised compensatory islet neogenesis. Additional evidence that a restriction of diet as a fetus and neonate can alter the phenotype of β-cell progenitor cells has arisen from experimentation *in vitro* with isolated islets. Islets were isolated from neonatal mice that had previously experienced a control or LP maternal diet. Over a 4 week culture period in dishes coated with a matrix of type 1 collagen the islets were found to de-differentiate when the media contained epidermal growth factor, giving rise to monolayers of cells with a ductal epithelial appearance [33]. Cells derived from offspring that received LP diet demonstrated a significant reduction in proliferation rate. If the cell monolayers were re-cultured in dishes coated with a Matrigel matrix in the presence of fibroblast growth factor-7 and insulin-like growth factor-II (IGF-II) they were found to form pseudo-islets within 4 weeks. Cells isolated from mice born to mothers fed a LP diet had a lower capacity to form pseudo-islets and demonstrated a lower insulin content and release. When the expression of transcription factors known to be involved in β-cell generation from progenitors was determined, pseudo-islets derived from control-fed offspring exhibited transcription factor expression close to that of freshly isolated islets. However, pseudo-islets derived from offspring exposed to LP diet failed to express Pdx1, neurogenin 3 and other key transcription factors, as well as insulin or somatostatin. Since β-cell neogenesis from endocrine precursors remains possible, albeit at low levels, throughout life in rodents [34], a depletion or functional deficit of this progenitor reserve in animals exposed to LP in early life may prevent adaptive changes in β-cell mass when facing metabolic stress, as is associated with obesity.

ANIMAL MODELS OF FETAL PROGRAMMING OF THE ENDOCRINE PANCREAS THROUGH HYPERGLYCEMIA

There are few rodent models of maternal hyperglycemia and/or diabetes that reproduce the pathophysiology of human pregnancy well, particularly with respect to the development of large birth weight, macrosomic newborn. However, the metabolic outcomes for the offspring following maturity are consistent with observations in human. Rat models typically render females diabetic by injection of STZ and allow several days for this to be cleared before subsequent mating [35, 36]. Offspring of diabetic mothers became glucose intolerant at 15 weeks of age [35]. The relative size of pancreatic islets was unaltered, but glucose-sensitive insulin secretion was relatively impaired. This resulted from intrinsic changes within the β-cells since the findings were reproducible with isolated islets, and a reduced expression of genes associated with β-cell glucose metabolism were found such as pyruvate dehydrogenase and pyruvate carboxylase. We have shown using transgenic mouse models that deletion of pyruvate dehydrogenase activity causes a delay in the development of the endocrine pancreas *in utero*, as well as impairing the glucose sensing by β-cells and insulin release [37]. Aref *et al.* [36] found that HOMA-insulin resistance was already greater in offspring of diabetic mothers at birth, as was HOMA- β-cell function. In offspring of diabetic rats, as in humans, a relative hyperinsulinemia existed at birth but resolved after the first two weeks of life. Islets were hypertrophied at birth but the large islets degenerated quickly after parturition resulting in a population of smaller islets than was seen in control offspring. An alternate model of fetal metabolic programming has been to feed the mother a high fat, high sucrose diet following conception [38]. The adult offspring were relatively obese with increased circulating leptin levels, demonstrated insulin resistance, and were relatively glucose intolerant; but changes in islet morphometry or function of the β-cells were not reported. A number of non-human primate models have also been utilized to model the immediate and long-lasting effects of diabetes during pregnancy on the metabolism of the offspring. A Western-style high fat diet was fed to macaques for several years before the females were mated. The offspring developed insulin resistance, a chronic inflammatory response resembling that seen in type 2 diabetes, islet hypertrophy and a relative change in the ratio of pancreatic β-cells to α-cells [39]. Islet vascularization was also impaired in the offspring, which is likely to result in reduced insulin secretion [40]. A comparable diet during human pregnancy resulted in maternal hyperglycemia, placental dysfunction and elevated levels of intra-hepatocellular lipid deposition of the fetus *in utero* [41].

It can be concluded that animal models of both nutritional insufficiency or imbalance during pregnancy, and the fetal hyperglycemia that accompanies pregnancies associated with maternal obesity and/or diabetes, both result in programmed changes in metabolic homeostasis in the offspring which are long-lasting and are markers of pre-diabetes in a human context. In each case the phenotype includes changes in the morphology or functional capacity of the islets of Langerhans. It will next be considered if there are common underlying cellular and molecular mechanisms that could underpin these changes.

MECHANISMS OF FETAL PROGRAMMING OF THE PANCREATIC FUNCTION

Early Dietary Insult and the mTOR Axis

The mammalian target of rapamycin (mTOR) is a serine/threonine kinase that links nutritional and environmental signals with those initiated by growth factors and hormones, resulting in a coordination of cell proliferation, cell hypertrophy, protein translation and metabolism [42 - 46]. Functional mTOR exists as two functional complexes, the mTOR Complex 1 (mTORC1) and mTORC2. The protein RAPTOR is an intrinsic part of mTORC1 whilst RICTOR is a necessary component of mTORC2 (Fig. **1**). Activation of mTORC1 by growth factors or amino acids results in a phosphorylation of ribosomal S6 kinase1 (S6K1) and an activation of the eukaryote initiation factor 4E-binding protein1 (4E-BP1), resulting in increased protein translation. Activation of mTORC2 phosphorylates PKCα and serum and glucocorticoid-regulated kinase 1 (SGK1) and is associated with cell proliferation [47, 48]. Thus the integrated actions of mTORC1 and 2 can control both cell number and function. A much-used tool to investigate the role of mTOR is rapamycin, which acts as a specific and rapid inhibitor of mTORC 1, but can also inhibit mTORC2 after prolonged exposure at high doses [49, 50]. There are also inhibitors of mTOR, the tuberous sclerosis complex (TSC) genes. Following the formation of a complex between TSC1 and TSC2 a phosphorylation of TSC2 by Akt or glycogen synthase kinase 3β (GSK3β) can occur, resulting in an activation of mTOR [51].

The mTOR pathway is intimately linked to β-cell mass and function. Rapamycin inhibited β-cell proliferation *in vitro* [52, 53] and prevented the expansion of the β-cell mass that is normally seen during pregnancy in the mouse [54], acting through inhibition of mTORC1. Rapamycin also prevented adaptation of an increased pancreatic β-cell mass in response to persistent hyperglycemia [55]. Mice deficient in functional S6K1 have decreased β-cell mass and hyperglycemia, demonstrating the importance of down stream signals of mTOR actions [56, 57]. Conversely, activation of mTOR by a deletion of the *TSC2* gene in β-cells resulted in cell proliferation and hypertrophy [58, 59], and improved glucose tolerance. The ability of rapamycin to reverse the effects of *TSC2* ablation would imply that β-cell proliferation was predominantly linked to mTORC1 signaling. However, *Rictor* null mice exhibited reduced β-cell mass, proliferation, glucose-stimulated insulin release and pancreas insulin content, suggesting that mTORC2 signaling is also involved in β-cell plasticity [60]. Signaling through mTOR also maintains β-cell survival, since the actions of insulin-like growth factor-I (IGF-I) in suppressing β-cell apoptosis through Akt is mediated by mTOR signaling [61]. Likewise, glucagon-like peptide-1 (GLP-1) promotes β-cell survival *via* activation of cyclic AMP, resulting in the mobilization of intracellular Ca^{2+}, which also leads to activation of mTOR [62].

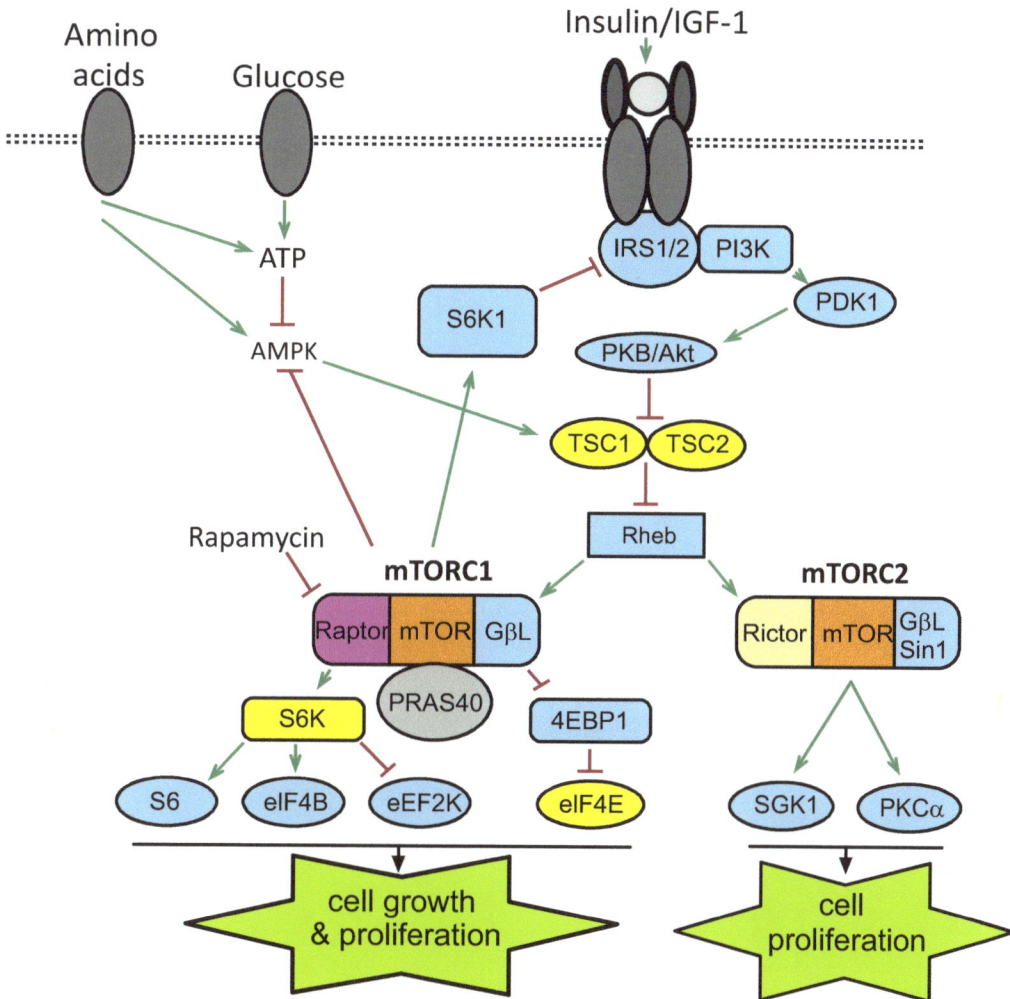

Fig. (1). Simplified mTOR signaling pathway showing relationships to β-cell proliferation and function. *Abbreviations*: key elements; mTORC1 and mTORC2 - mammalian target of rapamycin complex 1 and 2; IRS - insulin regulated substrate; PI3K - phosphatidylinositol 3-kinase; TSC1/2 - tuberous sclerosis complex1/2; 4E-BP1 - eukaryotic initiation factor binding protein; S6K - p70 S6 kinase; S6 - ribosomal protein S6; PKCα - protein kinase Cα; SGK1 - Serum and Glucocorticoid-regulated Kinase 1; eIF4B - eukaryotic translation initiation factor 4B; eEF2k - eukaryotic elongation factor-2 kinase; PDK1 - pyruvate dehydrogenase kinase 1; PKB/Akt - protein kinase B/AKT kinase; PRAS40 - proline-rich AKT Substrate of 40 kDa; ATP - adenosine triphosphate; AMPK - adenosine monophosphate - activated protein kinase.

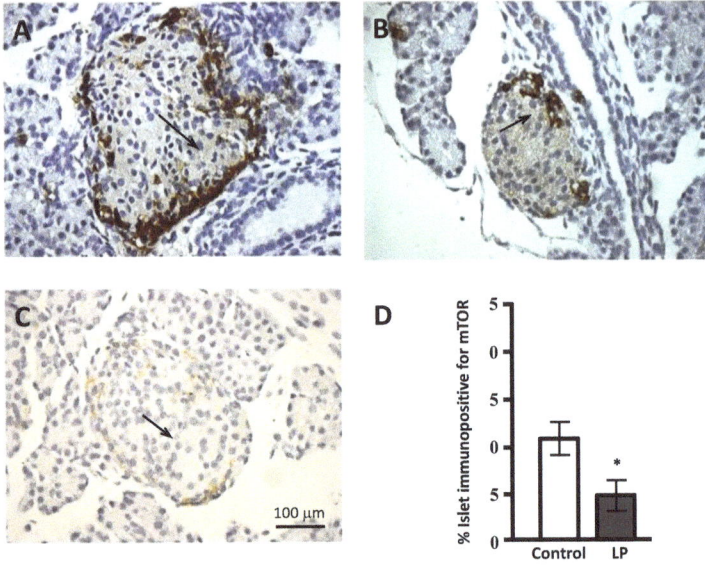

Fig. (2a). Immunohistochemical localization of mTOR in islets in 7 day old mice exposed to control [A & B] or low protein (LP) diet, and D) changes in the mean islet area occupied by mTOR-positive cells between diets. mTOR is present throughout the islets (arrows) in both α- and β-cells, but was significantly reduced in islets from mice exposed to LP diet, Mean±SEM; $n = 6$, *p<0.05 *vs* control diet.

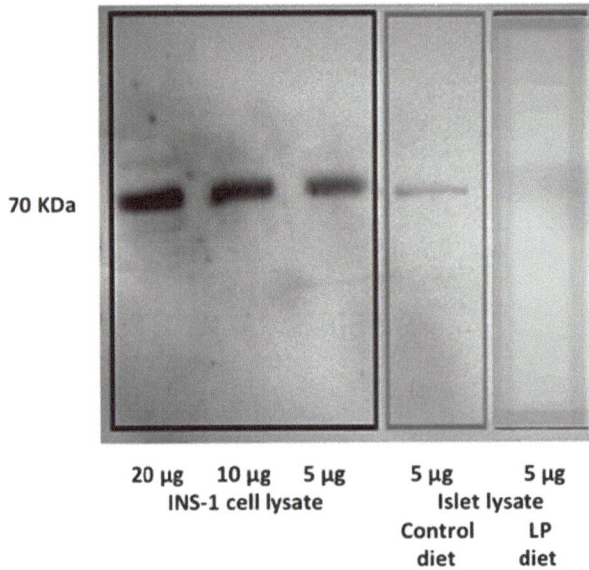

Fig. (2b). Western imunoblot analysis of p70 S6K protein in islet cell lysates from 7 day old offspring of pregnant mice fed control or LP diet, as well as INS-1 β-cell lysates as a control. p70 S6K was reduced in islets from LP-exposed neonatal mice.

The administration of a LP diet to young rats caused a decrease in the islet content of mTOR protein and glucose and amino acid-stimulated insulin release, and this was reversed by supplementation with leucine [63]. Similar changes in mTOR were found in the LP-treated pregnant rat in other tissues [64], whilst in mice a transient LP diet caused a fall and subsequent recovery of mTOR target kinases and β-cell mass, with the recovery being blocked by rapamycin [65]. In islets from offspring of LP-fed rats the observed decrease in nutrient-stimulated insulin release was accompanied by a reduced activation of the mTORC1 target, S6K1 [66]. Not surprisingly, β-cell mass is also dependent of protein synthesis since β-cells from mice where the S6K1 gene had been deleted in a cell-specific manner exhibited reduced β-cell size, number and insulin content [67]. We and others have examined the expression of the mTOR pathway in offspring from pregnant mice fed LP diet throughout gestation, with or without subsequent administration of STZ. Gene expression in whole pancreas or isolated islets from LP fed animals showed a substantially reduced expression of *mTOR* and cell cycle genes such as *cdk2* after STZ, compared to control-fed animals and coincident with a failure of β-cell regeneration [68, 69]. *TSC2* expression was significantly increased, but not so *Raptor* or *Rictor* mRNAs. Correspondingly, the presence of mTOR was much reduced in islets from LP-fed offspring when assessed by either western blot analysis or immunohistochemistry and as would be expected, the levels of p70 S6K1 were relatively lower in offspring receiving LP diet (Fig. **2**). Incubation of isolated islets from offspring of control-fed mice with rapamycin decreased the rate of β-cell proliferation to levels seen in islets isolated from LP-exposed offspring and increased the rate of apoptosis. This demonstrates that the functional implications of preventing mTOR signaling on islet regeneration involve deficits in β-cell proliferation, survival and function, and that a restriction in dietary protein in early life most likely causes changes to β-cell plasticity in later life through mechanisms involving long-lasting changes to mTOR signaling.

Epigenetic Mechanisms Governing Pancreatic Gene Expression and Their Involvement in Fetal Programming of Metabolic Disease

Substantial evidence is accumulating that epigenetic modifications to genomic DNA or to the supporting chromatin in response to environmental or nutritional insults *in utero* can have life long effects on the expression levels of genes involved with metabolic control in the offspring. The epigenetic marks most frequently encountered are modifications to DNA methylation, and altered acetylation or methylation of lysine residues in the histone regions of chromatin [70, 71]. The ability to have altered methylation of DNA on cytosine results from the activity levels of methyltransferase enzymes, and most frequently occurs within CG-rich sequences within gene promoter regions. Such sites are differentially methylated in over half of human genes, and typically an absence, or limited methylation of promoter regions results in a greater expression of the encoded proteins, whilst increased methylation results in a decrease expression. The amino terminal of histone proteins can be modified by acetylation, methylation or other chemical modifications, with a resulting change in the degree of openness or compaction of the DNA-chromatin complexes affecting the ability of transcription factors to interact with gene promoter regions. Increased acetylation of histones typically leads to increased gene transcription. Dietary manipulation *in utero* can alter DNA methylation in gene promoter regions, as shown by supplementation of maternal diet with methyl donors, such as folic acid, in the Agouti mouse [72]. Similarly, administration of LP diet to the pregnant rat resulted in promoter hypomethylation of the PPARα gene and the glucocorticoid receptor gene in liver of offspring, resulting in long-term changes to glucocorticoid sensitivity and altered metabolism [73, 74]. In a human context, adults who were malnourished *in utero* during the Dutch hunger winter of 1944, and who subsequently had a higher incidence of type 2 diabetes, had a decrease in methylation of the IGF-2 gene promoter region [75]. Conversely, exposure to diabetes *in utero* was associated with an altered methylation status for genes involved with glucose metabolism and others associated with cell growth when cells were harvested from cord blood or from placenta [76].

Specific epigenetic changes to genes involved with pancreas development *in utero* and β-cell function have been reported in response to nutritional manipulation in early life. This may commence during pancreatic embryogenesis. The pancreas develops from foregut endoderm, which derives from the definition of the germ layers and the differentiation of definitive endoderm. Pancreatic endoderm formation is controlled by a reduction of methylation on H3K27 histone, which suppresses stem cell pluripotency [77]. Subsequent pancreatic endoderm specification is dependent on the expression of transcription factors that include Foxa2 and Gata4 which cause acetylation changes on H3K27. Exposure to LP diet as early as blastocyst implantation can alter the fate map of the early embryo, and presumably this includes the presumptive pancreatic endoderm [78]. During organogenesis of the pancreas and the emergence of a normal population of functional β-cells, epigenetic manipulation through nutrition can modify the expression of key controlling transcription factors. Pdx1 is necessary for the definition of both the endocrine and exocrine pancreas from foregut endoderm, and subsequently is obligatory for the differentiation of β-cells and for insulin biosynthesis. When intrauterine growth retardation was induced in the rat levels of Pdx1 mRNA in the fetal pancreas were reduced substantially within 24h [79]. This did not recover following birth and the offspring became glucose intolerant as adults. Islets isolated from such growth-retarded fetuses exhibited a reduction in acetylation levels of H3 and H4 histones around the Pdx1 promoter, which resulted in an inability of the promoter region to be activated by transcription factors such as USF1 [80]. Following birth the epigenetic marks affecting Pdx1 continue to evolve with a progressive reduction in H3 acetylation but increased methylation of H3K9, and a progressive reduction in Pdx1 expression with age resulting in deficient β-cell mass and impaired glucose-stimulated insulin secretion. Pdx1 can regulate mitochondrial transcription factor A (Tfam) directly, which in turn can regulate the replication of mitochondrial DNA and is also reduced in islets from animals that exhibited IUGR [81]. The deficits in mitochondrial function can result in increased oxidative stress and a greater risk of β-cell damage during hyperglycemia. A second key gene susceptible to epigenetic modification is hepatocyte nuclear factor 4α (HNF-4α), which is required for β-cell development *in utero* and postnatally. Rats born to mothers exposed to LP diet during gestation exhibited a reduced interaction between the P2 promoter, which is active in pancreas, and the HNF-4α enhancer region, resulting in a decreased gene expression [82]. Acetylation at H3 and methylation at H3K4 were decreased whilst methylation at H3K9 was increased. These changes also occurred with age, but were prematurely advanced following LP diet. An additional variable is the epigenetic plasticity that exists between the α-cell and β-cell phenotypes, such that a balance between methylation of H3K4 and H3K27 can result in β-cells becoming α-cells [83], although it is not known how this might be altered specifically by fetal nutrition. While epigenetic programming of β-cell mass and function are likely to contribute to a postnatal risk of diabetes it is most likely that these occur in parallel with long-lasting changes in insulin sensitivity and glucose transport in peripheral tissues, through epigenetic modulation of the glucose transporter, Glut4 [84], and that it is the combination that leads to metabolic disease.

Epigenetic Changes to Pancreatic Gene Expression Caused by Environmental Toxins

A number of environmental toxins have been shown to contribute to fetal programming of postnatal diabetes risk, and at least some of these alter pancreatic islet function. These include arsenic and other heavy metals that cause hypomethylation of DNA in CpG islands, mitochondrial oxidative stress, and glucose metabolism [85]. Similarly, cadmium exposure caused changes in DNA methyltransferase and DNA methylation patterns [86]. Nicotine exposure to the fetus through maternal smoking can alter uteroplacental efficiency causing a reduction in β-cell mass in the offspring with life-long deficiency [87]. Recently, attention has focused on endocrine disruptors, such as bisphenol A (BPA), a phytoestrogen and a common contaminant in plastics, which is known to modify epigenetic marks in a variety of tissues [88]. Mice treated with BPA during gestation became glucose intolerant and insulin insensitive following parturition for up to seven months, with a reduced β-cell mass and glucose-stimulated insulin release *in vitro* [89]. The β-cell effects included a reduced cell proliferation rate with a lowered cyclin D1 and cdk4, higher levels of the β-cell cycle inhibitor, p16, and a greater incidence of cell apoptosis. In a similar model male offspring from both the F1 and F2 generations demonstrated abnormal glucose tolerance and increased body fat following exposure *in utero* to BPA [90]. This was associated with altered DNA methylation of the IGF-2 gene. Since BPA has been detected in cord blood it is highly likely that the human fetal can be similarly affected [91].

Prematurity of Cellular Aging

While it has been argued that the observed changes in tissue phenotype and function in the offspring as a result of an adverse gestational environment are predictive-adaptive mechanisms in anticipation of an expected postnatal nutritional environment [92], they might also represent more general aspects of premature cellular aging, as typified by insulin resistance, increased cellular oxidative stress, and a less adaptive β-cell mass following metabolic challenge. Shortening of telomere length within chromosomes is a fundamental feature of cellular aging and is controlled by the RNA-dependent DNA polymerase, telomerase. The telomerase core enzyme includes telomerase reverse transcriptase (TERT) and the RNA template (TERC). In the human fetus developmental shortening of telomeres is dynamic, with telomere length decreasing rapidly between 6 and 7 weeks gestation, and then declining only slightly until birth [93]. An inverse correlation was seen between TERT or TERC expression and gestational age. Psychological and physiological stress has been associated with an increased rate of telomere shortening throughout life [94], and the same appears to be true at a cellular level. In the placenta from infants with intrauterine growth restriction a reduced copy number of the TERC gene was reported, leading to decreased telomerase and enhanced telomere shortening [95]. Similar observations were made from cord blood mononuclear cells from growth-restricted term infants [96]. TERT exists in mitochondria as well as the nucleus, and is increased in response to the cellular oxidative stress that occurs in offspring of diabetic mothers when measured in cord blood mononuclear cells [97]. This may protect cells from oxidative stress during gestation, but the altered ratio of TERT within mitochondria *vs.* nuclei may result in an increased rate of cellular aging in the longer term, resulting in premature changes to β-cell survival and function, and in insulin sensitivity. Cherif *et al.* [98] reported that the relative abundance of shortened telomeres was increased with aging in rat tissues, which included the pancreas, kidney, liver, and lung. This resulted in premature cell senescence, but the brain was spared. The offspring of rats receiving LP diet during pregnancy had accelerated telomere shortening in pancreatic islets and many other tissues compared to offspring of control-fed mothers, and in islets this was accompanied by increased levels of p16[INK], an inhibitor of β-cell proliferation [99]. The lifespan of the offspring from mothers who received LP diet was shortened [100], supporting the concept that premature tissue aging is a contributor to the risk of adult metabolic disease.

REVERSAL STRATEGIES FOR FETAL PROGRAMMING OF THE PANCREAS

Two available windows of intervention to prevent of reverse fetal programming of adult diabetes would be during gestation in at-risk individuals, or in the neonatal period. The former is only safely amenable to maternal nutritional supplementation. The β-cell phenotype in the offspring of rats fed a LP diet during pregnancy was rescued if the diet was supplemented with taurine [17, 18]. This included a protection of β-cell mass, normalization of cell cycle kinetics, appropriate glucose-stimulated insulin release and a reduction in islet oxidative stress. The deficits in intra-islet vascular volume and VEGF receptor abundance seen in offspring of LP-fed rats at birth were also reversed by supplementation with taurine, and glucose tolerance was normal when the offspring reached adulthood. Taurine is localized to both α- and δ-cells within the islets [101], and is the most substantially reduced of all the amino acids within the fetal serum following exposure to maternal LP diet [102, 103]. While adult humans can maintain appropriate endogenous taurine levels from dietary methionine and cysteine, the synthetic capacity is much lower in the neonate, and the deficiency is amplified in infants with intrauterine growth restriction [104]. Similarly, nutritional supplementation of mothers with methyl donors such as folic acid, or micronutrient-rich vegetables may also improve newborn outcome [105].

In other models of intrauterine growth retardation the reductions in β-cell mass and islet vasculature present at birth can be reversed in neonatal life by administration of anabolic peptide drugs such as exendin, a GLP-1 analog [106]. Exendin 4 treatment increased histone acetylase activity and reversed the epigenetic pattern occurring on Pdx1 gene expression in IUGR offspring [107].

CONCLUSION

The balance of evidence suggests that the future structure and function of the pancreatic β-cells can be altered by the intra-uterine environment both in the formative period of embryogenesis and during subsequent tissue maturation prior to birth. An adverse environment can exist due to an imbalance in maternal nutrition, through the fetal hyperglycemia and hyperlipidemia that can occur during poorly-controlled diabetes during pregnancy, sub-optimal utero-placental function as can result from maternal infection or smoking, or the effects of environmental endocrine disruptors such as BPA. The trajectory of performance of the endocrine pancreas can be set inappropriately for the offspring stretching into adulthood, but may only become manifest as metabolic disease following a subsequent challenge, such as the β-cell oxidative stress associated with postnatal obesity, or the metabolic stress of pregnancy. In such situations the normal adaptive mechanisms to increase β-cell mass and insulin release are rendered sub-optimal due to changes in cellular pathways initiated *in utero*. These include changes in the ability to generate new β-cells from a progenitor pool within the pancreas, adaptive changes in β-cell mass through the integrative actions of the mTOR signaling pathway, long-term changes in the expression levels of key genes controlling β-cell proliferation and insulin biosynthesis as determined by epigenetic modifications, and a possible change in the processes of cellular aging. However, evidence that nutritional supplementation of the mother during pregnancy with micronutrients, or the treatment of the neonates with β-cell trophic peptides, can reverse long-term deficits in the plasticity of β-cell mass and function in the F1 generation suggests that these developmental 'clocks' can potentially be reset.

CONFLICT OF INTEREST

The author confirms that he has no conflict of interest to declare for this publication.

ACKNOWLEDGEMENTS

For cited studies published by the author he is grateful for research funding from the Alan Thicke Centre for Juvenile Diabetes Research, the Program of Experimental Medicine, Department of medicine, Western University, and the European Commission Framework 7 Program.

REFERENCES

[1]   Barker DJ. Programming the baby Mothers, babies, disease in later life. London: Br Med J Publishing Group 1994; pp. 14-36.

[2]   Painter RC, Roseboom TJ, Bleker OP. Prenatal exposure to the Dutch famine and disease in later life:

an overview. Reprod Toxicol 2005; 20(3): 345-52.
[http://dx.doi.org/10.1016/j.reprotox.2005.04.005] [PMID: 15893910]

[3]   Devlieger R, Casteels K, Van Assche FA. Reduced adaptation of the pancreatic B cells during pregnancy is the major causal factor for gestational diabetes: current knowledge and metabolic effects on the offspring. Acta Obstet Gynecol Scand 2008; 87(12): 1266-70.
[http://dx.doi.org/10.1080/00016340802443863] [PMID: 18846453]

[4]   Plagemann A. Maternal diabetes and perinatal programming. Early Hum Dev 2011; 87(11): 743-7.
[http://dx.doi.org/10.1016/j.earlhumdev.2011.08.018] [PMID: 21945359]

[5]   Clausen TD, Mathiesen ER, Hansen T, *et al*. High prevalence of type 2 diabetes and pre-diabetes in adult offspring of women with gestational diabetes mellitus or type 1 diabetes: the role of intrauterine hyperglycemia. Diabetes Care 2008; 31(2): 340-6.
[http://dx.doi.org/10.2337/dc07-1596] [PMID: 18000174]

[6]   Yessoufou A, Moutairou K. Maternal diabetes in pregnancy: early and long-term outcomes on the offspring and the concept of 'metabolic memory'. Diabetes Res 2011; 2011: 1-12.
[http://dx.doi.org/dx.doi.org/10.1155/2011/218598]

[7]   Langley-Evans SC. Nutrition in early life and the programming of adult disease: a review. J Hum Nutr Diet 2015; 28 (Suppl. 1): 1-14.
[http://dx.doi.org/10.1111/jhn.12212] [PMID: 24479490]

[8]   Cripps RL, Martin-Gronert MS, Archer ZA, Hales CN, Mercer JG, Ozanne SE. Programming of hypothalamic neuropeptide gene expression in rats by maternal dietary protein content during pregnancy and lactation. Clin Sci 2009; 117(2): 85-93.
[http://dx.doi.org/10.1042/CS20080393] [PMID: 19152506]

[9]   Mantzoros CS, Rifas-Shiman SL, Williams CJ, Fargnoli JL, Kelesidis T, Gillman MW. Cord blood leptin and adiponectin as predictors of adiposity in children at 3 years of age: a prospective cohort study. Pediatrics 2009; 123(2): 682-9.
[http://dx.doi.org/10.1542/peds.2008-0343] [PMID: 19171638]

[10]   Fowden AL, Forhead AJ. Glucocorticoids as regulatory signals during intrauterine development. Exp Physiol 2015; 100(12): 1477-87.
[http://dx.doi.org/10.1113/EP085212] [PMID: 26040783]

[11]   Hill DJ. Nutritional programming of pancreatic β-cell plasticity. World J Diabetes 2011; 2(8): 119-26.
[http://dx.doi.org/10.4239/wjd.v2.i8.119] [PMID: 21954415]

[12]   Martín MA, Alvarez C, Goya L, Portha B, Pascual-Leone AM. Insulin secretion in adult rats that had experienced different underfeeding patterns during their development. Am J Physiol 1997; 272(4 Pt 1): E634-40.
[PMID: 9142885]

[13]   Manuel-Apolinar L, Rocha L, Damasio L, Tesoro-Cruz E, Zarate A. Role of prenatal undernutrition in the expression of serotonin, dopamine and leptin receptors in adult mice: implications of food intake. Mol Med Rep 2014; 9(2): 407-12.
[PMID: 24337628]

[14]   Simmons RA, Templeton LJ, Gertz SJ. Intrauterine growth retardation leads to the development of

type 2 diabetes in the rat. Diabetes 2001; 50(10): 2279-86.
[http://dx.doi.org/10.2337/diabetes.50.10.2279] [PMID: 11574409]

[15]   Somm E, Schwitzgebel VM, Vauthay DM, *et al.* Prenatal nicotine exposure alters early pancreatic islet and adipose tissue development with consequences on the control of body weight and glucose metabolism later in life. Endocrinology 2008; 149(12): 6289-99.
[http://dx.doi.org/10.1210/en.2008-0361] [PMID: 18687784]

[16]   Petrik J, Reusens B, Arany E, *et al.* A low protein diet alters the balance of islet cell replication and apoptosis in the fetal and neonatal rat and is associated with a reduced pancreatic expression of insulin-like growth factor-II. Endocrinology 1999; 140(10): 4861-73.
[PMID: 10499546]

[17]   Boujendar S, Reusens B, Merezak S, *et al.* Taurine supplementation to a low protein diet during foetal and early postnatal life restores a normal proliferation and apoptosis of rat pancreatic islets. Diabetologia 2002; 45(6): 856-66.
[http://dx.doi.org/10.1007/s00125-002-0833-6] [PMID: 12107730]

[18]   Boujendar S, Arany E, Hill DJ, Remacle C, Reusens B. Taurine supplementation during fetal life reverses the vascular impairment caused to the endocrine pancreas by a low protein diet. J Nutr 2003; 133: 2820-5.
[PMID: 12949371]

[19]   Chamson-Reig A, Thyssen SM, Hill DJ, Arany E. Exposure of the pregnant rat to low protein diet causes impaired glucose homeostasis in the young adult offspring by different mechanisms in males and females. Exp Biol Med (Maywood) 2009; 234(12): 1425-36.
[http://dx.doi.org/10.3181/0902-RM-69] [PMID: 19657071]

[20]   Reusens B, Theys N, Dumortier O, Goosse K, Remacle C. Maternal malnutrition programs the endocrine pancreas in progeny. Am J Clin Nutr 2011; 94(6) (Suppl.): 1824S-9S.
[http://dx.doi.org/10.3945/ajcn.110.000729] [PMID: 21562089]

[21]   Rao S, Yajnik CS, Kanade A, *et al.* Intake of micronutrient-rich foods in rural Indian mothers is associated with the size of their babies at birth: Pune Maternal Nutrition Study. J Nutr 2001; 131(4): 1217-24.
[PMID: 11285330]

[22]   Fall C. Maternal nutrition: effects on health in the next generation. Indian J Med Res 2009; 130(5): 593-9.
[PMID: 20090113]

[23]   Nikolova G, Jabs N, Konstantinova I, *et al.* The vascular basement membrane: a niche for insulin gene expression and β cell proliferation. Dev Cell 2006; 10(3): 397-405.
[http://dx.doi.org/10.1016/j.devcel.2006.01.015] [PMID: 16516842]

[24]   Johansson M, Mattsson G, Andersson A, Jansson L, Carlsson P-O. Islet endothelial cells and pancreatic β-cell proliferation: studies *in vitro* and during pregnancy in adult rats. Endocrinology 2006; 147(5): 2315-24.
[http://dx.doi.org/10.1210/en.2005-0997] [PMID: 16439446]

[25]   Lammert E, Cleaver O, Melton D. Role of endothelial cells in early pancreas and liver development. Mech Dev 2003; 120(1): 59-64.

[http://dx.doi.org/10.1016/S0925-4773(02)00332-5] [PMID: 12490296]

[26]   Nicholson JM, Arany EJ, Hill DJ. Changes in islet microvasculature following streptozotocin-induced β-cell loss and subsequent replacement in the neonatal rat. Exp Biol Med (Maywood) 2010; 235(2): 189-98.
[http://dx.doi.org/10.1258/ebm.2009.009316] [PMID: 20404034]

[27]   Pladys P, Sennlaub F, Brault S, *et al.* Microvascular rarefaction and decreased angiogenesis in rats with fetal programming of hypertension associated with exposure to a low-protein diet in utero. Am J Physiol Regul Integr Comp Physiol 2005; 289(6): R1580-8.
[http://dx.doi.org/10.1152/ajpregu.00031.2005] [PMID: 16037123]

[28]   Ferreira RV, Gombar FM, da Silva Faria T, Costa WS, Sampaio FJ, da Fonte Ramos C. Metabolic programming of ovarian angiogenesis and folliculogenesis by maternal malnutrition during lactation. Fertil Steril 2010; 93(8): 2572-80.
[http://dx.doi.org/10.1016/j.fertnstert.2009.05.050] [PMID: 19591993]

[29]   Sathishkumar K, Elkins R, Yallampalli U, Yallampalli C. Protein restriction during pregnancy induces hypertension and impairs endothelium-dependent vascular function in adult female offspring. J Vasc Res 2009; 46(3): 229-39.
[http://dx.doi.org/10.1159/000166390] [PMID: 18957856]

[30]   Joanette EA, Reusens B, Arany E, Thyssen S, Remacle RC, Hill DJ. Low-protein diet during early life causes a reduction in the frequency of cells immunopositive for nestin and CD34 in both pancreatic ducts and islets in the rat. Endocrinology 2004; 145(6): 3004-13.
[http://dx.doi.org/10.1210/en.2003-0796] [PMID: 15044374]

[31]   Cox AR, Gottheil SK, Arany EJ, Hill DJ. The effects of low protein during gestation on mouse pancreatic development and β cell regeneration. Pediatr Res 2010; 68(1): 16-22.
[http://dx.doi.org/10.1203/PDR.0b013e3181e17c90] [PMID: 20386490]

[32]   Goosse K, Bouckenooghe T, Sisino G, Aurientis S, Remacle C, Reusens B. Increased susceptibility to streptozotocin and impeded regeneration capacity of beta-cells in adult offspring of malnourished rats. Acta Physiol (Oxf) 2014; 210(1): 99-109.
[http://dx.doi.org/10.1111/apha.12121] [PMID: 23701924]

[33]   Beamish CA, Strutt B, Arany E, Hill DJ. Imaging of re-differentiated islet like clusters using a novel immunocytochemical method. Diabetes 2007; 56 (Suppl. 1): PA425.

[34]   De Groef S, Leuckx G, Van Gassen N, *et al.* Surgical injury to the mouse pancreas through ligation of the pancreatic duct as a model for endocrine and exocrine reprogramming and proliferation. J Vis Exp 2015; 102(102): e52765.
[http://dx.doi.org/10.3791/52765] [PMID: 26273954]

[35]   Han J, Xu J, Long YS, Epstein PN, Liu YQ. Rat maternal diabetes impairs pancreatic β-cell function in the offspring. Am J Physiol Endocrinol Metab 2007; 293(1): E228-36.
[http://dx.doi.org/10.1152/ajpendo.00479.2006] [PMID: 17389712]

[36]   Aref AB, Ahmed OM, Ali LA, Semmler M. Maternal rat diabetes mellitus deleteriously affects insulin sensitivity and beta-cell function in the offspring. J Diabetes Res 2013; 429154.
[http://dx.doi.org/dx.doi.org/10.1155/2013/429154] [PMID: 429154]

[37]   Patel MS, Srinivasan M, Strutt B, Mahmood S, Hill DJ. Featured Article: Beta cell specific pyruvate dehydrogenase alpha gene deletion results in a reduced islet number and β-cell mass postnatally. Exp Biol Med (Maywood) 2014; 239(8): 975-85.
[http://dx.doi.org/10.1177/1535370214531895] [PMID: 24845368]

[38]   Zheng J, Xiao X, Zhang Q, *et al.* Maternal and post-weaning high-fat, high-sucrose diet modulates glucose homeostasis and hypothalamic POMC promoter methylation in mouse offspring. Metab Brain Dis 2015; 30(5): 1129-37.
[http://dx.doi.org/10.1007/s11011-015-9678-9] [PMID: 25936720]

[39]   Comstock SM, Pound LD, Bishop JM, *et al.* High-fat diet consumption during pregnancy and the early post-natal period leads to decreased a-cell plasticity in the nonhuman primate. Mol Med 2013; 2: 10-22.

[40]   Pound LD, Comstock SM, Grove KL. Consumption of a Western-style diet during pregnancy impairs offspring islet vascularization in a Japanese macaque model. Am J Physiol Endocrinol Metab 2014; 307(1): E115-23.
[http://dx.doi.org/10.1152/ajpendo.00131.2014] [PMID: 24844258]

[41]   Brumbaugh DE, Tearse P, Cree-Green M, *et al.* Intrahepatic fat is increased in the neonatal offspring of obese women with gestational diabetes. J Pediatr 2013; 162(5): 930-6.e1.
[http://dx.doi.org/10.1016/j.jpeds.2012.11.017] [PMID: 23260099]

[42]   Peng T, Golub TR, Sabatini DM. The immunosuppressant rapamycin mimics a starvation-like signal distinct from amino acid and glucose deprivation. Mol Cell Biol 2002; 22(15): 5575-84.
[http://dx.doi.org/10.1128/MCB.22.15.5575-5584.2002] [PMID: 12101249]

[43]   Tee AR, Blenis J. mTOR, translational control and human disease. Semin Cell Dev Biol 2005; 16(1): 29-37.
[http://dx.doi.org/10.1016/j.semcdb.2004.11.005] [PMID: 15659337]

[44]   Martin DE, Hall MN. The expanding TOR signaling network. Curr Opin Cell Biol 2005; 17(2): 158-66.
[http://dx.doi.org/10.1016/j.ceb.2005.02.008] [PMID: 15780592]

[45]   Hay N, Sonenberg N. Upstream and downstream of mTOR. Genes Dev 2004; 18(16): 1926-45.
[http://dx.doi.org/10.1101/gad.1212704] [PMID: 15314020]

[46]   Jacinto E, Hall MN. Tor signalling in bugs, brain and brawn. Nat Rev Mol Cell Biol 2003; 4(2): 117-26.
[http://dx.doi.org/10.1038/nrm1018] [PMID: 12563289]

[47]   Balcazar N, Sathyamurthy A, Elghazi L, *et al.* mTORC1 activation regulates β-cell mass and proliferation by modulation of cyclin D2 synthesis and stability. J Biol Chem 2009; 284(12): 7832-42.
[http://dx.doi.org/10.1074/jbc.M807458200] [PMID: 19144649]

[48]   Jacinto E, Loewith R, Schmidt A, *et al.* Mammalian TOR complex 2 controls the actin cytoskeleton and is rapamycin insensitive. Nat Cell Biol 2004; 6(11): 1122-8.
[http://dx.doi.org/10.1038/ncb1183] [PMID: 15467718]

[49]   Sarbassov DD, Ali SM, Sengupta S, *et al.* Prolonged rapamycin treatment inhibits mTORC2 assembly and Akt/PKB. Mol Cell 2006; 22(2): 159-68.

[http://dx.doi.org/10.1016/j.molcel.2006.03.029] [PMID: 16603397]

[50]   Rosner M, Hengstschläger M. Cytoplasmic and nuclear distribution of the protein complexes mTORC1 and mTORC2: rapamycin triggers dephosphorylation and delocalization of the mTORC2 components rictor and sin1. Hum Mol Genet 2008; 17(19): 2934-48.
[http://dx.doi.org/10.1093/hmg/ddn192] [PMID: 18614546]

[51]   Inoki K, Li Y, Xu T, Guan KL. Rheb GTPase is a direct target of TSC2 GAP activity and regulates mTOR signaling. Genes Dev 2003; 17(15): 1829-34.
[http://dx.doi.org/10.1101/gad.1110003] [PMID: 12869586]

[52]   Nir T, Melton DA, Dor Y. Recovery from diabetes in mice by beta cell regeneration. J Clin Invest 2007; 117(9): 2553-61.
[http://dx.doi.org/10.1172/JCI32959] [PMID: 17786244]

[53]   Niclauss N, Bosco D, Morel P, Giovannoni L, Berney T, Parnaud G. Rapamycin impairs proliferation of transplanted islet β cells. Transplantation 2011; 91(7): 714-22.
[PMID: 21297554]

[54]   Zahr E, Molano RD, Pileggi A, *et al.* Rapamycin impairs beta-cell proliferation *in vivo.* Transplant Proc 2008; 40(2): 436-7.
[http://dx.doi.org/10.1016/j.transproceed.2008.02.011] [PMID: 18374093]

[55]   Fraenkel M, Ketzinel-Gilad M, Ariav Y, *et al.* mTOR inhibition by rapamycin prevents beta-cell adaptation to hyperglycemia and exacerbates the metabolic state in type 2 diabetes. Diabetes 2008; 57(4): 945-57.
[http://dx.doi.org/10.2337/db07-0922] [PMID: 18174523]

[56]   Pende M, Kozma SC, Jaquet M, *et al.* Hypoinsulinaemia, glucose intolerance and diminished beta-cell size in S6K1-deficient mice. Nature 2000; 408(6815): 994-7.
[http://dx.doi.org/10.1038/35050135] [PMID: 11140689]

[57]   Jastrzebski K, Hannan KM, Tchoubrieva EB, Hannan RD, Pearson RB. Coordinate regulation of ribosome biogenesis and function by the ribosomal protein S6 kinase, a key mediator of mTOR function. Growth Factors 2007; 25(4): 209-26.
[http://dx.doi.org/10.1080/08977190701779101] [PMID: 18092230]

[58]   Rachdi L, Balcazar N, Osorio-Duque F, *et al.* Disruption of Tsc2 in pancreatic β cells induces β cell mass expansion and improved glucose tolerance in a TORC1-dependent manner. Proc Natl Acad Sci USA 2008; 105(27): 9250-5.
[http://dx.doi.org/10.1073/pnas.0803047105] [PMID: 18587048]

[59]   Bartolomé A, Guillén C, Benito M. Role of the TSC1-TSC2 complex in the integration of insulin and glucose signaling involved in pancreatic β-cell proliferation. Endocrinology 2010; 151(7): 3084-94.
[http://dx.doi.org/10.1210/en.2010-0048] [PMID: 20427478]

[60]   Gu Y, Lindner J, Kumar A, Yuan W, Magnuson MA. Rictor/mTORC2 is essential for maintaining a balance between β-cell proliferation and cell size. Diabetes 2011; 60(3): 827-37.
[http://dx.doi.org/10.2337/db10-1194] [PMID: 21266327]

[61]   Cai Y, Wang Q, Ling Z, *et al.* Akt activation protects pancreatic beta cells from AMPK-mediated death through stimulation of mTOR. Biochem Pharmacol 2008; 75(10): 1981-93.

[http://dx.doi.org/10.1016/j.bcp.2008.02.019] [PMID: 18377870]

[62]　Kwon G, Marshall CA, Pappan KL, Remedi MS, McDaniel ML. Signaling elements involved in the metabolic regulation of mTOR by nutrients, incretins, and growth factors in islets. Diabetes 2004; 53 (Suppl. 3): S225-32.
[http://dx.doi.org/10.2337/diabetes.53.suppl_3.S225] [PMID: 15561916]

[63]　Filiputti E, Rafacho A, Araújo EP, *et al.* Augmentation of insulin secretion by leucine supplementation in malnourished rats: possible involvement of the phosphatidylinositol 3-phosphate kinase/mammalian target protein of rapamycin pathway. Metabolism 2010; 59(5): 635-44.
[http://dx.doi.org/10.1016/j.metabol.2009.09.007] [PMID: 19913855]

[64]　Rosario FJ, Jansson N, Kanai Y, Prasad PD, Powell TL, Jansson T. Maternal protein restriction in the rat inhibits placental insulin, mTOR, and STAT3 signaling and down-regulates placental amino acid transporters. Endocrinology 2011; 152(3): 1119-29.
[http://dx.doi.org/10.1210/en.2010-1153] [PMID: 21285325]

[65]　Crozier SJ, DAlecy LG, Ernst SA, Ginsburg LE, Williams JA. Molecular mechanisms of pancreatic dysfunction induced by protein malnutrition. Gastroenterology 2009; 137(3): 1093-1101, 1101.e1-1101.e3.
[http://dx.doi.org/10.1053/j.gastro.2009.04.058] [PMID: 19427311]

[66]　Filiputti E, Ferreira F, Souza KL, *et al.* Impaired insulin secretion and decreased expression of the nutritionally responsive ribosomal kinase protein S6K-1 in pancreatic islets from malnourished rats. Life Sci 2008; 82(9-10): 542-8.
[http://dx.doi.org/10.1016/j.lfs.2007.12.012] [PMID: 18234235]

[67]　Um SH, Sticker-Jantscheff M, Chau GC, *et al.* S6K1 controls pancreatic β cell size independently of intrauterine growth restriction. J Clin Invest 2015; 125(7): 2736-47.
[http://dx.doi.org/10.1172/JCI77030] [PMID: 26075820]

[68]　Cox AR, Beamish CA, Carter DE, Arany EJ, Hill DJ. Cellular mechanisms underlying failed beta cell regeneration in offspring of protein-restricted pregnant mice. Exp Biol Med (Maywood) 2013; 238(10): 1147-59.
[PMID: 23986224]

[69]　Alejandro EU, Gregg B, Wallen T, *et al.* Maternal diet-induced microRNAs and mTOR underlie β cell dysfunction in offspring. J Clin Invest 2014; 124(10): 4395-410.
[http://dx.doi.org/10.1172/JCI74237] [PMID: 25180600]

[70]　Fuks F. DNA methylation and histone modifications: teaming up to silence genes. Curr Opin Genet Dev 2005; 15(5): 490-5.
[http://dx.doi.org/10.1016/j.gde.2005.08.002] [PMID: 16098738]

[71]　Simmons RA. Developmental origins of β-cell failure in type 2 diabetes: the role of epigenetic mechanisms. Pediatr Res 2007; 61(5 Pt 2): 64R-7R.
[http://dx.doi.org/10.1203/pdr.0b013e3180457623] [PMID: 17413845]

[72]　Cooney CA, Dave AA, Wolff GL. Maternal methyl supplements in mice affect epigenetic variation and DNA methylation of offspring. J Nutr 2002; 132(8) (Suppl.): 2393S-400S.
[PMID: 12163699]

[73]   Lillycrop KA, Slater-Jefferies JL, Hanson MA, Godfrey KM, Jackson AA, Burdge GC. Induction of altered epigenetic regulation of the hepatic glucocorticoid receptor in the offspring of rats fed a protein-restricted diet during pregnancy suggests that reduced DNA methyltransferase-1 expression is involved in impaired DNA methylation and changes in histone modifications. Br J Nutr 2007; 97(6): 1064-73.
[http://dx.doi.org/10.1017/S000711450769196X] [PMID: 17433129]

[74]   Lillycrop KA, Phillips ES, Torrens C, Hanson MA, Jackson AA, Burdge GC. Feeding pregnant rats a protein-restricted diet persistently alters the methylation of specific cytosines in the hepatic PPAR alpha promoter of the offspring. Br J Nutr 2008; 100(2): 278-82.
[http://dx.doi.org/10.1017/S0007114507894438] [PMID: 18186951]

[75]   Heijmans BT, Tobi EW, Stein AD, et al. Persistent epigenetic differences associated with prenatal exposure to famine in humans. Proc Natl Acad Sci USA 2008; 105(44): 17046-9.
[http://dx.doi.org/10.1073/pnas.0806560105] [PMID: 18955703]

[76]   El Hajj N, Pliushch G, Schneider E, et al. Metabolic programming of MEST DNA methylation by intrauterine exposure to gestational diabetes mellitus. Diabetes 2013; 62(4): 1320-8.
[http://dx.doi.org/10.2337/db12-0289] [PMID: 23209187]

[77]   Xie R, Everett LJ, Lim HW, et al. Dynamic chromatin remodeling mediated by polycomb proteins orchestrates pancreatic differentiation of human embryonic stem cells. Cell Stem Cell 2013; 12(2): 224-37.
[http://dx.doi.org/10.1016/j.stem.2012.11.023] [PMID: 23318056]

[78]   Fleming TP, Velazquez MA, Eckert JJ. Embryos, DOHaD and David Barker. J Dev Orig Health Dis 2015; 6(5): 377-83. Epub ahead of print
[http://dx.doi.org/10.1017/S2040174415001105] [PMID: 25952250]

[79]   Pinney SE, Simmons RA. Metabolic programming, epigenetics, and gestational diabetes mellitus. Curr Diab Rep 2012; 12(1): 67-74.
[http://dx.doi.org/10.1007/s11892-011-0248-1] [PMID: 22127642]

[80]   Park JH, Stoffers DA, Nicholls RD, Simmons RA. Development of type 2 diabetes following intrauterine growth retardation in rats is associated with progressive epigenetic silencing of Pdx1. J Clin Invest 2008; 118(6): 2316-24.
[PMID: 18464933]

[81]   Gauthier BR, Wiederkehr A, Baquié M, et al. PDX1 deficiency causes mitochondrial dysfunction and defective insulin secretion through TFAM suppression. Cell Metab 2009; 10(2): 110-8.
[http://dx.doi.org/10.1016/j.cmet.2009.07.002] [PMID: 19656489]

[82]   Sandovici I, Smith NH, Nitert MD, et al. Maternal diet and aging alter the epigenetic control of a promoter-enhancer interaction at the Hnf4a gene in rat pancreatic islets. Proc Natl Acad Sci USA 2011; 108(13): 5449-54.
[http://dx.doi.org/10.1073/pnas.1019007108] [PMID: 21385945]

[83]   Bramswig NC, Everett LJ, Schug J, et al. Epigenomic plasticity enables human pancreatic α to β cell reprogramming. J Clin Invest 2013; 123(3): 1275-84.
[http://dx.doi.org/10.1172/JCI66514] [PMID: 23434589]

[84]   Raychaudhuri N, Raychaudhuri S, Thamotharan M, Devaskar SU. Histone code modifications repress glucose transporter 4 expression in the intrauterine growth-restricted offspring. J Biol Chem 2008; 283(20): 13611-26.
[http://dx.doi.org/10.1074/jbc.M800128200] [PMID: 18326493]

[85]   Xie Y, Liu J, Benbrahim-Tallaa L, *et al.* Aberrant DNA methylation and gene expression in livers of newborn mice transplacentally exposed to a hepatocarcinogenic dose of inorganic arsenic. Toxicology 2007; 236(1-2): 7-15.
[http://dx.doi.org/10.1016/j.tox.2007.03.021] [PMID: 17451858]

[86]   Takiguchi M, Achanzar WE, Qu W, Li G, Waalkes MP. Effects of cadmium on DNA-(Cytosine-5) methyltransferase activity and DNA methylation status during cadmium-induced cellular transformation. Exp Cell Res 2003; 286(2): 355-65.
[http://dx.doi.org/10.1016/S0014-4827(03)00062-4] [PMID: 12749863]

[87]   Somm E. Nicotinic cholinergic signaling in adipose tissue and pancreatic islets biology: revisited function and therapeutic perspectives. Arch Immunol Ther Exp (Warsz) 2014; 62(2): 87-101.
[http://dx.doi.org/10.1007/s00005-013-0266-6] [PMID: 24276790]

[88]   Ho SM, Tang WY, Belmonte de Frausto J, Prins GS. Developmental exposure to estradiol and bisphenol A increases susceptibility to prostate carcinogenesis and epigenetically regulates phosphodiesterase type 4 variant 4. Cancer Res 2006; 66(11): 5624-32.
[http://dx.doi.org/10.1158/0008-5472.CAN-06-0516] [PMID: 16740699]

[89]   Alonso-Magdalena P, García-Arévalo M, Quesada I, Nadal Á. Bisphenol-A treatment during pregnancy in mice: a new window of susceptibility for the development of diabetes in mothers later in life. Endocrinology 2015; 156(5): 1659-70.
[http://dx.doi.org/10.1210/en.2014-1952] [PMID: 25830705]

[90]   Susiarjo M, Xin F, Bansal A, *et al.* Bisphenol a exposure disrupts metabolic health across multiple generations in the mouse. Endocrinology 2015; 156(6): 2049-58.
[http://dx.doi.org/10.1210/en.2014-2027] [PMID: 25807043]

[91]   Vandenberg LN, Chahoud I, Heindel JJ, Padmanabhan V, Paumgartten FJ, Schoenfelder G. Urinary, circulating, and tissue biomonitoring studies indicate widespread exposure to bisphenol A. Environ Health Perspect 2010; 118(8): 1055-70.
[http://dx.doi.org/10.1289/ehp.0901716] [PMID: 20338858]

[92]   Gluckman PD, Hanson MA, Spencer HG. Predictive adaptive responses and human evolution. Trends Ecol Evol (Amst) 2005; 20(10): 527-33.
[http://dx.doi.org/10.1016/j.tree.2005.08.001] [PMID: 16701430]

[93]   Cheng G, Kong F, Luan Y, *et al.* Differential shortening rate of telomere length in the development of human fetus. Biochem Biophys Res Commun 2013; 442(1-2): 112-5.
[http://dx.doi.org/10.1016/j.bbrc.2013.11.022] [PMID: 24246679]

[94]   Shalev I, Entringer S, Wadhwa PD, *et al.* Stress and telomere biology: a lifespan perspective. Psychoneuroendocrinology 2013; 38(9): 1835-42.
[http://dx.doi.org/10.1016/j.psyneuen.2013.03.010] [PMID: 23639252]

[95]   Biron-Shental T, Kidron D, Sukenik-Halevy R, *et al.* TERC telomerase subunit gene copy number in

placentas from pregnancies complicated with intrauterine growth restriction. Early Hum Dev 2011; 87(2): 73-5.
[http://dx.doi.org/10.1016/j.earlhumdev.2010.08.024] [PMID: 21168289]

[96]   Davy P, Nagata M, Bullard P, Fogelson NS, Allsopp R. Fetal growth restriction is associated with accelerated telomere shortening and increased expression of cell senescence markers in the placenta. Placenta 2009; 30(6): 539-42.
[http://dx.doi.org/10.1016/j.placenta.2009.03.005] [PMID: 19359039]

[97]   Li P, Tong Y, Yang H, *et al.* Mitochondrial translocation of human telomerase reverse transcriptase in cord blood mononuclear cells of newborns with gestational diabetes mellitus mothers. Diabetes Res Clin Pract 2014; 103(2): 310-8.
[http://dx.doi.org/10.1016/j.diabres.2013.12.024] [PMID: 24480248]

[98]   Cherif H, Tarry JL, Ozanne SE, Hales CN. Ageing and telomeres: a study into organ- and gender-specific telomere shortening. Nucleic Acids Res 2003; 31(5): 1576-83.
[http://dx.doi.org/10.1093/nar/gkg208] [PMID: 12595567]

[99]   Tarry-Adkins JL, Martin-Gronert MS, Fernandez-Twinn DS, *et al.* Poor maternal nutrition followed by accelerated postnatal growth leads to alterations in DNA damage and repair, oxidative and nitrosative stress, and oxidative defense capacity in rat heart. FASEB J 2013; 27(1): 379-90.
[http://dx.doi.org/10.1096/fj.12-218685] [PMID: 23024373]

[100] Jennings BJ, Ozanne SE, Dorling MW, Hales CN. Early growth determines longevity in male rats and may be related to telomere shortening in the kidney. FEBS Lett 1999; 448(1): 4-8.
[http://dx.doi.org/10.1016/S0014-5793(99)00336-1] [PMID: 10217398]

[101] Bustamante J, Lobo MV, Alonso FJ, *et al.* An osmotic-sensitive taurine pool is localized in rat pancreatic islet cells containing glucagon and somatostatin. Am J Physiol Endocrinol Metab 2001; 281(6): E1275-85.
[PMID: 11701444]

[102] Reusens B, Dahri S, Snoeck A, Bennis Taleb N, Remacle C, Hoet JJ. Long-term consequences of diabetes and its complications may have a fetal origin: experimental and epidemiological evidence. New York: Raven Press 1995; Vol. 35: pp. 187-98.

[103] Cherif H, Reusens B, Ahn M-T, Hoet JJ, Remacle C. Effects of taurine on the insulin secretion of rat fetal islets from dams fed a low-protein diet. J Endocrinol 1998; 159(2): 341-8.
[http://dx.doi.org/10.1677/joe.0.1590341] [PMID: 9795376]

[104] Zelikovic I, Chesney RW, Friedman AL, Ahlfors CE. Taurine depletion in very low birth weight infants receiving prolonged total parenteral nutrition: role of renal immaturity. J Pediatr 1990; 116(2): 301-6.
[http://dx.doi.org/10.1016/S0022-3476(05)82898-7] [PMID: 2105387]

[105] Potdar RD, Sahariah SA, Gandhi M, *et al.* Improving womens diet quality preconceptionally and during gestation: effects on birth weight and prevalence of low birth weighta randomized controlled efficacy trial in India (Mumbai Maternal Nutrition Project). Am J Clin Nutr 2014; 100(5): 1257-68.
[http://dx.doi.org/10.3945/ajcn.114.084921] [PMID: 25332324]

[106] Ham JN, Crutchlow MF, Desai BM, Simmons RA, Stoffers DA. Exendin-4 normalizes islet vascularity in intrauterine growth restricted rats: potential role of VEGF. Pediatr Res 2009; 66(1): 42-

6.
[http://dx.doi.org/10.1203/PDR.0b013e3181a282a5] [PMID: 19287346]

[107] Pinney SE, Jaeckle Santos LJ, Han Y, Stoffers DA, Simmons RA. Exendin-4 increases histone acetylase activity and reverses epigenetic modifications that silence Pdx1 in the intrauterine growth retarded rat. Diabetologia 2011; 54(10): 2606-14.
[http://dx.doi.org/10.1007/s00125-011-2250-1] [PMID: 21779870]

CHAPTER 5

Malprogramming of β-Cell Function by a Dietary Modification in the Immediate Postnatal Period

Mulchand S. Patel* and **Saleh Mahmood**

Department of Biochemistry, School of Medicine and Biomedical Sciences, University at Buffalo, The State University of New York, Buffalo, New York 14214, USA

Abstract: The development of the structure and function of the endocrine pancreas is known to be influenced by altered nutritional experience during the fetal period. Nutritional modifications in the suckling period are also recognized as contributing factors to developmental programming of the endocrine pancreas. In this chapter we describe the malprogramming of rat pancreatic islet structure and β cell functions in response to an increased intake of carbohydrate-derived calories in a milk formula (HC) during the suckling period. Alterations in β cell function of HC rat pups result in the development of hyperinsulinemia due to β cell plasticity in the immediate postnatal period. These modifications include: altered islet architecture and increased insulin-producing mass, increased insulin secretion capacity with a leftward shift in glucose-stimulated insulin secretion, insulin secretion in the absence of glucose and/or Ca^{2+}, increased gene transcription of several genes crucial for β cell development and function, and increased parasympathetic input, as well as malprogramming of orexigenic circuitry in the hypothalamus. Interestingly, these alterations in β cell function are maintained even after weaning of HC rats on a standard rodent chow, resulting in adult-onset obesity due to development of hyperphagia. It is possible that early introduction of carbohydrate-rich infant supplemental foods could contribute to modified β cell functions in infants which could, in turn, over a longer period predispose to the development of childhood obesity and/or adult-onset obesity and its associated metabolic complications including type 2 diabetes.

* **Corresponding author Mulchand S. Patel:** Department of Biochemistry, School of Medicine and Biomedical Sciences, University at Buffalo, Buffalo, NY 14214, USA; Tel: 716-829-3074, Fax: 716-829-2725; E-mail: mspatel@buffalo.edu

David J. Hill (Ed.)

Keywords: Artificial rearing of rat pups, β cell neogenesis, β cell proliferation, Hyperinsulinemia, Hypothalamic programming, Insulin secretion, Metabolic programming, Nutritional modification, Obesity, *Pdx-1* gene expression.

INTRODUCTION

Metabolic Programming After Birth

Numerous lines of evidence from animal models, epidemiological studies and clinical investigations indicate that altered nutritional exposures during critical periods of early development (fetal and immediate postnatal periods) can have permanent effects at the cellular, molecular, and biochemical levels in several tissues [1 - 7]. During this rapid growth period, the endocrine pancreas in rodents undergoes both normal structural and biochemical maturation of the endocrine functions [8, 9]. Altered nutrition during these early phases of life can induce permanent adaptations in these processes which continue to be expressed in adulthood [6, 10 - 14]. This phenomenon was initially proposed by Barker [15] as 'Fetal Origins of Adult Disease' followed by other similar terms and more recently referred to as 'Developmental Origins of Health and Disease'. Altered nutritional experiences during the fetal period can be exerted due to maternal malnutrition (such as caloric restriction, protein deficiency, calorie plus protein deficiency, vitamin deficiency), maternal obesity and maternal diabetes.

Accumulated evidence indicates that nutritional alterations (undernutrition, overnutrition and altered milk composition) in the immediate postnatal period (referred to as the suckling period) alone can malprogram metabolic capacities of tissues (*e.g.* pancreatic β cells and hypothalamus) with increased risk for the development of metabolic disorders in adulthood. Human milk is the natural choice for infants because it not only provides nutrients for optimal growth but is also the source of bioactive compounds and immunoglobulins for immunity development during the immediate postnatal period [16]. Commercially available infant formulas are devoid of the latter components. The health benefits of breast-feeding over infant formula feeding are widely recognized, including optimal growth during childhood and longer term benefits such as lower rates of cardiovascular risk and obesity in adulthood [16, 17]. During the suckling period, over-nourishment may be caused by infant milk-formula feeding (bottle feeding with unrestricted supply of milk formula) with or without early introduction of carbohydrate-enriched infant foods. Although exclusive breast-feeding of babies for the first six months is recommended by the American Dietetic Association [18], formula feeding (aka bottle feeding) and early introduction of complementary infant foods are widely practiced in Westernized societies. It is quite possible that feeding formula alone may result in over-feeding causing increased weight gain and possible metabolic programming of infants. Rodent models of overnutrition by intake of excess maternal milk during the suckling period (*e.g.* by reducing the litter size) are well characterized for metabolic programming of the hypothalamus and pancreatic β cells, resulting in excess weight gain during the immediate postnatal period and predisposition for the development of adult on-set obesity [19 - 22].

Unlike human milk and infant milk formula with a higher level of calories derived from fat, supplementary infant foods (fruits, juices, cereals) are highly enriched with carbohydrate-derived calories but have few calories from fats. Hence, early introduction of infant foods prior to 6 months of age can increase carbohydrate-derived calorie ingestion. The focus of our research has been on metabolic programming effects in β cells due to increased intake of carbohydrate *via* a milk-substitute formula by rat pups during the suckling period [23 - 25]. To investigate this phenomenon we have employed a rat model (referred to as Pup-in-a-Cup model) in which rat pups are artificially reared by intra-gastric feeding of a milk formula high in carbohydrate-derived calories during the suckling period. Based on the observations on the HC rat model, as presented in this chapter, it is possible that early introduction of sugar-dense supplementary foods for human infants could be a contributing factor to metabolic programming of β cells and hence in the etiology of obesity in childhood and adulthood.

EXPERIMENTAL APPROACHES: 'PUP-IN-A-CUP' RAT MODEL

To feed a modified rat milk-substitute formula enriched in carbohydrate-derived calories to neonates during the suckling period, we took advantage of an artificial rearing technique employing intragastric feeding of rats [26 - 28]. On postnatal day 4, pups born to normal dams consuming standard laboratory chow *ad libitum* throughout their lives, were randomly assigned to the control or experimental group. The control group (mother-fed; MF) consisted of pups reared naturally (*i.e.* consuming rat milk) by dams. The caloric composition of rat milk was 8% carbohydrate, 68% fat, and 24% protein. The experimental group (HC) consisted of artificially reared pups on a high carbohydrate milk formula (caloric composition: 56% carbohydrate, 20% fats, and 24% protein). To rule out any adverse effect of the artificial rearing technique, another group of neonatal pups (referred to as HF) were artificially reared and fed a high-fat milk formula with the caloric distribution similar to that of rat milk (8% carbohydrate, 68% fat, and 24% protein) [28]. Artificially reared pups with intragastric cannulas were individually housed in Styrofoam cups floating in a 37°C water bath (hence referred to as Pup-in-a-Cup model). The details of this technique are reported elsewhere [23, 29]. It should be emphasized that the outcome of several parameters for HF rats (which are reported in several publications) were not different compared with that of MF rats, clearly supporting that any alterations observed in HC rats were due to the HC dietary modification and not due to employment of an artificial rearing technique in our investigations [26, 30].

Alterations in Islets Mass

In rodents, rapid increases in islets mass occur in late fetal and neonatal periods, resulting from a balance between islet cell neogenesis, replication and programmed death. These processes are shown to be very sensitive to nutritional insults during critical periods of development. During the suckling period, rodent milk rich in fat-derived calorie and low in carbohydrate-derived calories supports normal programmed development of pancreatic endocrine cell growth, maturation and function, and prepares the pancreas to respond to a dietary switch to a carbohydrate-rich chow diet during the weaning process. In our experimental model of HC rat feeding a milk formula high in carbohydrate-derived calories by gastrostomy, this switch to a diet high in carbohydrate-derived calories occurs early on postnatal day 4 compared to normal weaning on day 21, impacting normal development of the pancreatic endocrine cellular and molecular processes, resulting in development of hyperinsulinemia in HC pups.

To analyze adaptive changes at the cellular level, islet ontogeny in HC pups during the sucking period (up to 24 days) was investigated [31]. Islet number per unit area was significantly increased in HC pups from the postnatal day 6 to 24 (Fig. **1A**). The size distribution between small and large islets revealed a greater percentage of small islets in HC pups, many of them being small endocrine cell clusters with 6 to 20 cells only (Fig. **1B**). Interestingly, islets from HC pups had a significantly reduced relative area occupied by glucagon-positive α cells at age 2 weeks, but an increased area occupied by insulin-positive β cells, resulting in a significantly increased β/α cell ratio (Table **1**). Islets from HC pups had an increased incidence of apoptosis with increased fragmented DNA. In HC pups, there was increased immunopositive cell staining for cell replication in ductal epithelium, a source of new islets by neogenesis, with lower incidence of apoptosis, supporting an overall increase in islet mass [31]. Expression of insulin-like growth factor II (IGF-II) mRNA in whole pancreas and IGF-II immunostaining in islets cells were reduced in HC pups. In contrast, the percentage of IGF-II positive ductal epithelium cells was significantly increased in HC pups. Altered IGF-II expression may be responsible, in part, for modified pancreatic islet ontogeny with altered islet size and number in HC pups.

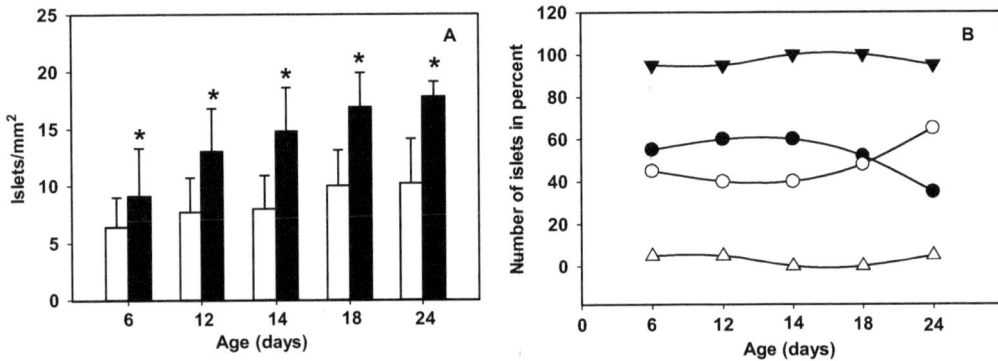

Fig. (1). Changes with age in the number of islets per square millimeter of pancreas (A) and percent distribution of islets according to size (small *vs* large) in pancreatic histologic sections from MF and HC rats during postnatal age 6 to 24 days. (A): open bar, MF; closed bar, HC. (B): ●, MF small islets; ○, MF large islets; ▼, HC small islets; △, HC large islets. * P<0.05. Data were taken from Petrik *et al.* [31].

Molecular Adaptations

Nutrient-mediated regulation of gene expression is an important mechanism by which an organism adapts to an altered nutritional environment during development. Significant increases were observed in the level of preproinsulin mRNA and also for insulin biosynthesis measured as incorporation of ^{3}H-leucine in newly synthesized insulin in the islets from 12-day-old HC rats (Table **1**) [32]. Pancreatic duodenal homeobox factor-1 (*Pdx-1*) plays a critical role not only in pancreatic organogenesis by committing the pluripotent epithelial cells to endocrine lineage, but also as a key transactivator of preproinsulin gene transcription. The mRNA levels, protein content and DNA-binding activity of Pdx-1 were all significantly increased in HC islets, supporting increased levels of insulin biosynthesis and storage in HC islets. Additionally, the mRNA levels of the stress-activated protein kinase-2, phosphatidylinositol 3-kinase, upstream stimulatory factor-1, Isl-1 and REGIII were also significantly increased in islets from HC pups [32]. These gene products are involved in upregulation of *Pdx-1* levels and function. Also, the mRNA levels of *IRS-1*, *IRS-2*, glucose transporter 2, and acetyl-CoA carboxylase which are necessary for intracellular signaling and generation of intermediates involved in insulin secretion by β cells, were also increased in islets from HC pups [33].

Table 1. Alterations in the structure and/or function of tissues and in serum hormonal levels in HC rat pups.

Tissue	Changes in parameters
Pancreas	Increased number of islets Increased size of islets Increased β/α ratio in islets Increased GSIS Leftward shift in GSIS Increased gene transcription Increased insulin synthesis Increased insulin stores Increased GLP-1R mRNA
Hypothalamus	Increased orexigenic neuropeptides Decreased anorexigenic neuropeptides Increased parasympathetic activity Decreased sympathetic activity
Serum	Increased insulin levels Increased GLP-1 levels Decreased leptin levels

For adaptive responses in cellular and biochemical processes of the cell, the changes in transcriptional activity of a large number of genes are necessary to support the altered phenotype of the β cell. A greater than 2-fold change in gene expression for a large number of genes was observed in islets from HC pups and adult HC rats [33]. These upregulated genes belonged to several specific function clusters such as cell cycle regulators, DNA-binding and chromatin proteins, transcriptional factors, oncogenes and tumor suppressor. Collectively, the genes from these clusters could contribute to alterations in ontogenesis of the endocrine pancreas, contributing to changes in the size and number of islets in the pancreas of HC pups. Furthermore, a wide array of increased gene expression was observed in support of the increased insulin secretory process (Table **1**). The clusters of genes implicated in this process and upregulated due to nutritional intervention included receptors, modulators, effectors and intracellular transducers, ion channel and transport proteins, extracellular cell signaling, communication and metabolic pathways. Collectively, these altered molecular events occurring during the critical period in pancreatic organogenesis in response to high carbohydrate intake during the suckling period play a critical role for adaptive changes at the cellular and biochemical levels in islets from HC rats.

MOLECULAR MECHANISM BY WHICH GLUCOSE CAN CONTROL β-CELL MASS

Soon after birth, the endocrine pancreas in rodents undergoes remodeling involving well-coordinated processes for β cell proliferation and apoptosis [9]. These processes appear to be altered in the endocrine pancreas of HC pups by an increased number of smaller size islets with increased β/α cell ratio. Immediate onset of hyperinsulinemia within 24 hours in HC pups represents an initial, primary response to dietary glucose (Fig. **2**) [26]. This is supported by the absence of increased serum insulin in pups fed a milk formula high in fat-derived calories intragastrically. Glucose metabolism is shown to have positive effects on gene transcription in β cells especially on the *Pdx-1* and preproinsulin genes [34 - 37]. Additionally, glucose promotes the translocation of *Pdx-1* protein from the cytoplasm to the nucleus for preproinsulin gene transcription [38]. The importance of glucose metabolism for β cell growth and maturation was shown in mouse models with β cell-specific deficiency in glucokinase [39] and pyruvate dehydrogenase complex activities [40]. In the glucokinase-deficient mouse model, both β cell proliferation as well as cell survival was impaired and glucose-stimulated insulin secretion (GSIS) was abolished [39]. In the pyruvate dehydrogenase complex-deficient mouse, the islet density as well as β cell mass was markedly reduced during the immediate postnatal period [40].

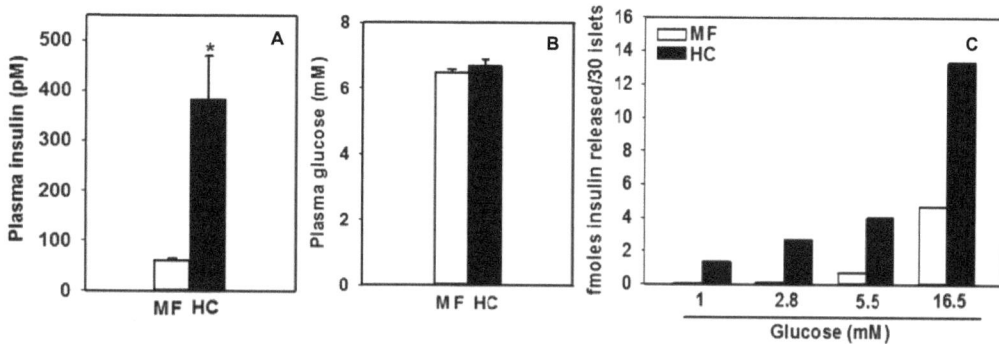

Fig. (2). Plasma levels of insulin (A) and glucose (B) as well as glucose-stimulated insulin secretion by isolated islets at 60 min (C) from 12 day-old MF (open bar) and HC (closed bar) rats. * P<0.05. Data are taken from Aalinkeel *et al.* [41].

Early on-set hyperinsulinemia in HC pups may result in autocrine effects of the insulin signaling pathway on β cell proliferation to support increased mass. Insulin acting through IRS-2 in the insulin signaling pathway can regulate β cell growth and proliferation [42 - 45]. This is achieved by increasing *Pdx-1* expression [46]. Similarly, HC islet hyperinsulinemia was also accompanied by increased *Pdx-1* expression. Another possible contributing factor to β cell proliferation is a gut incretin, GLP-1 (Fig. **2**). GLP-1 increases β cell proliferation and this effect are dependent on *Pax4* expression [44, 47]. Hence, increased levels of serum GLP-1 as well as increased levels of GLP-1R mRNA in β cells from HC pups could contribute to β cell proliferation (Fig. **3**), (Table **1**).

Alterations in Nutrient-Mediated Insulin Secretion

Insulin secretion by pancreatic β cells is regulated by both nutrient and non-nutrient secretagogues, with glucose being the primary and most important physiological stimulus. The insulinotropic action of glucose is exerted by its metabolism to increase the cytosolic ATP: ADP ratio as well as generation of several metabolites involved in metabolism-insulin secretion coupling in β cells [49, 50]. Several metabolites generated from glucose-derived pyruvate metabolism in the mitochondria (generating ATP, GTP, and glutamate) and subsequently from citrate metabolism in the cytosol (producing malonyl-CoA, NADPH) serve as mediators of GSIS [51 - 54]. An increase in the cytosolic ATP:ADP ratio results in the following sequential events in β cells: closure of K_{ATP} channels, depolarization of the plasma membrane, opening of voltage-sensitive L-type Ca^{2+} channels, Ca^{2+} influx, and insulin exocytosis [49, 50].

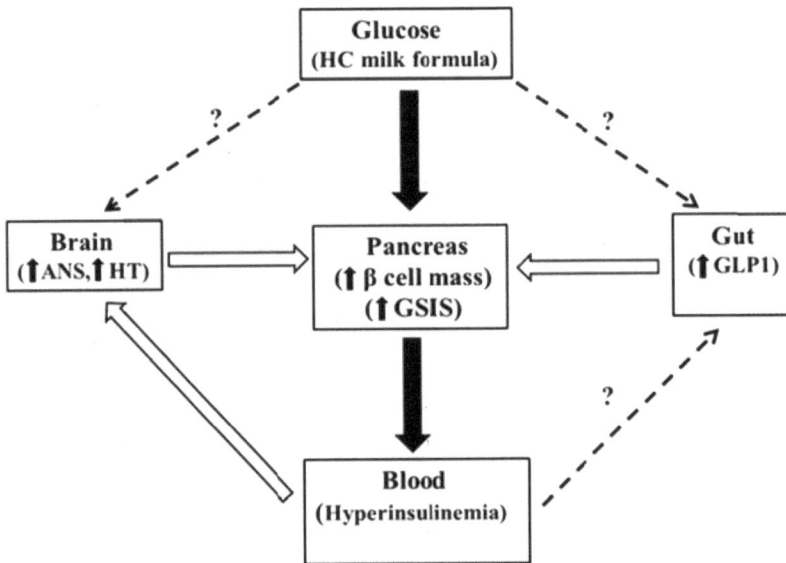

Fig. (3). Possible mechanisms involved in programming of hyperinsulinemia in HC rat pups due to feeding of a milk formula high in carbohydrate-derived calorie during the immediate postnatal period. GSIS, glucose-stimulated insulin secretion; ANS, Autonomic nervous system; HT, hypothalamus; GLP1, glucagon-like polypeptide 1. This figure is adapted from Patel and Srinivasan [48].

Fetal rodent islets respond poorly to GSIS. They, however, acquire functional maturation for GSIS around the second week of postnatal life with improvement in hormonal secretory function beyond the suckling period [55 - 58]. In the HC rat model, the high-carbohydrate milk formula treatment was initiated on postnatal day 4 and HC pups immediately developed hyperinsulinemia (~5-fold increase) but maintained normoglycemia (Fig. **2A, B**) [26, 41]. Pancreatic content of insulin was also significantly increased (~1.5-fold) in HC pups. GSIS patterns from HC rat pup islets at varying concentrations of glucose were markedly different compared with islets from control pups. As expected, islets from control pups secreted no detectable or very low amounts of insulin at sub-basal levels of glucose (1 and 2.8 mM) and secreted a low level of insulin at a basal level of glucose (5 mM) at 60 min (Fig. **2C**). In contrast, islets from HC rat pups secreted significantly higher amounts of insulin even at sub-basal glucose concentrations, and showed robust insulin secretory response at 60 min with 5 mM and 16.7 mM glucose levels (Fig. **2C**); (Table **1**) [41, 59]. Interestingly, these results show the lowering of the threshold for GSIS (the leftward shift), making it possible for HC islets to secrete insulin at a basal plasma glucose level for maintenance of sustained hyperinsulinemia in HC pups.

For insulin secretion by islets, transport and metabolism of glucose are prerequisite for the generation of cytoplasmic ATP and several coupling metabolites for insulin exocytosis. Significant increases in glucose transporter-2 protein content and in the activities of cytoplasmic hexokinase, glucokinase, glyceraldehyde-3-phosphate dehydrogenase (glycolytic enzymes) and mitochondrial pyruvate dehydrogenase complex (both active and total activities) were observed in islets from HC rat pups (Fig. **4**) [41]. Increased hexokinase activity in HC rat islets coincides with their increased ability to secret insulin at basal and even at sub-basal levels of glucose. Increased activities of these cytoplasmic and mitochondrial enzymes also support the enhanced capacity of HC rat islets for insulin secretion at higher glucose concentrations. As shown in Table **2**, it is of interest to note that HC rat islets were able to secret insulin in measurable amounts (i) in the absence of added glucose, (ii) in the presence of the non-metabolized 2-deoxyglucose, and (iii) when glycolysis was inhibited by the presence of iodoacetate. Furthermore, mannoheptulose (11 mM), a glucokinase inhibitor, had no effect on insulin secretion by both MF and HC islets at 5.5 mM glucose but at 16.7 mM glucose insulin secretion was reduced to basal levels in either MF and HC islets [41]. Collectively, these findings clearly indicate that insulin secretion under basal conditions is markedly increased in islets from HC rat pups by a novel mechanism(s) that remains unclear.

Table 2. Insulin secretory response by isolated islets from 12 day-old MF and HC rat pups to modulators of insulin secretion. Data are taken from Aalinkeel *et al.* [41].

Conditions	Glucose (mM)	Additions/deletions	MF	HC
			Insulin release (fmol/30 islets/h)	
Substrate metabolism	0	None	N.D.	1.6
	0	2-deoxyglucose (11 mM)	N.D.	1.8
	5.5	Iodoacetate (1mM)	N.D.	1.8
Glucokinase inhibition	5.5	None	0.45	3.9
	5.5	Mannoheptulose (11 mM)	0.42	4.2
	16.7	None	4.7	13.4
	16.7	Mannoheptulose (11 mM)	1.4	4.0
Calcium deprivation	5.5	Nimidipine (1 μM)	N.D.	1.6
	5.5	BAPTA (1 μM)	N.D.	1.7
	5.5	EGTA (500 μM)	N.D.	1.7
Potassium channel blockade	5.5	KCl (25 mM)	2.9	3.4
	5.5	Glibenclamide (0.1 mM)	3.6	3.0

There were also several unusual characteristics of HC islets with regard to their ability to secrete insulin in the presence or absence of specific mediators (*e.g.* Ca^{2+}). For example, in contrast to MF islets, HC islets secreted modest amounts of insulin in the absence of added extracellular Ca^{2+} (*i.e.*, Ca^{2+}-free buffer) or depleted intracellular Ca^{2+} stores or in the presence of blocked voltage-gated Ca^{2+} channels (by addition of 1μM nimodipine or 1 μM BAPTA) (Table **2**) [41]. The ability of HC islets to secrete modest amounts of insulin even in the absence of added extracellular Ca^{2+} or in the presence of blocked voltage-gated Ca^{2+} channels raised two possibilities: (i) operation of a Ca^{2+}-independent insulin secretion pathway (Fig. **4**), and/or (ii) as yet undetermined structural alteration(s) in β cells allowing insulin secretion in the absence of Ca^{2+}. Furthermore, when K^+ channels were blocked and the cells depolarized (in the presence of 100 μM glibenclamide or 25 mM potassium chloride), no effect was detected on insulin secretion by HC islets whereas similarly treated MF islets secreted significantly higher levels of insulin compared to their basal levels without a treatment (Table **2**) [41]. Interestingly, these findings indicated that HC islets were able to secrete moderate amounts of insulin through a glucose- and calcium-independent mechanism that does not involve glucose metabolism nor increased levels of intracellular Ca^{2+} (Fig. **4**).

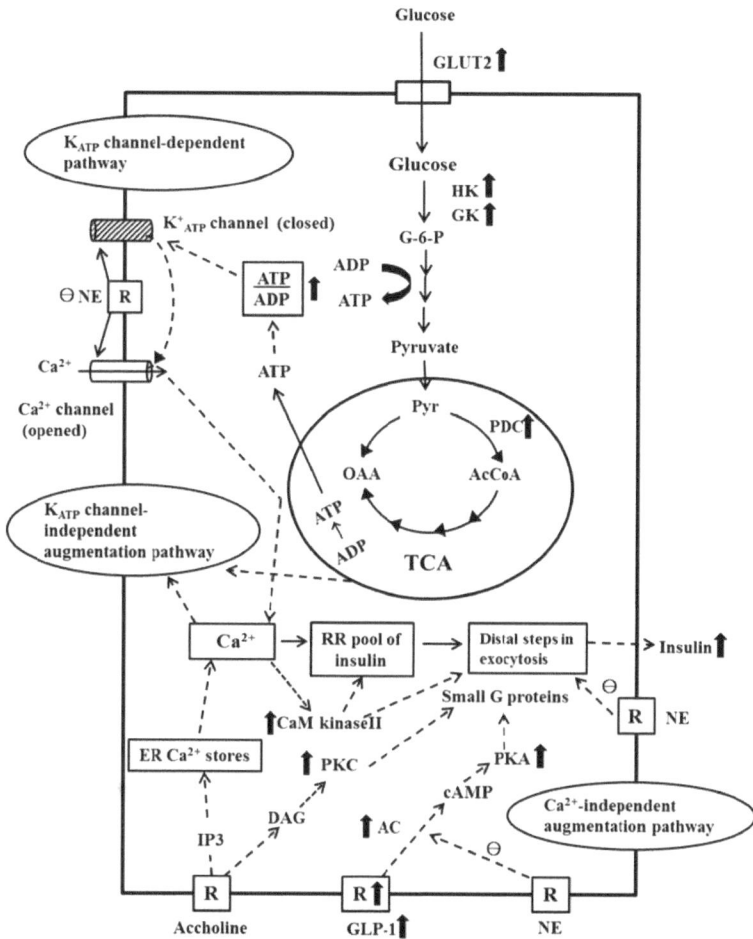

Fig. (4). Summary of the signaling pathways altered by hormonal and neuronal inputs in islets isolated from HC rat pups subjected to the high-carbohydrate milk formula feeding. R, receptor; ER, endoplasmic reticulum; NE, norepinephrine; IP3, inositol 1, 4, 5-triphosphote; GLUT2, glucose transporter 2; HK, hexokinase; GK, glucokinase; PDC, pyruvate dehydrogenase complex; AcCoA, acetyl-CoA; OA, oxaloacetate; TCA, tricarboxylic acid cycle; AE, adenylate cyclase; Accholine, acetylcholine; RR, readily releasable; DAG, diacylglycerol; ⊖ , reduced sensitivity to NE-sensitive pathways; ↑, indicates increase in activity of mRNA content. This figure is adapted from Srinivasan *et al.* [60].

Alterations in Non-Nutrient-Mediated Insulin Secretion

In addition to nutrient-mediated insulin secretion, hormones and neural inputs also modulate insulin secretion by β cells. Increases in plasma GLP-1 levels and in the level of GLP-1 receptor mRNA were detected in islets of HC pups [60]. Both in the presence (5 mM) or absence of glucose, GLP-1 (*via* activation of cAMP production and PKA activation) and acetylcholine independently stimulated insulin secretion by HC islets compared to MF islets. An activity of 'active' form of PKA activity without any significant change in 'total' PKA activity as well as activities of PKC and CaM kinase II was significantly increased in HC islets [60]. Since increases in the levels of intracellular Ca^{2+}, cAMP and diacylglycerol activate CaM kinase II, PKA, and PKC, respectively, these findings support increased capacity for insulin secretion by HC islets (Fig. **4**). Of five adenylyl cyclase isoforms (II, III, V, VI, and VII) present in rat islets, isoform VII was found to be most highly expressed isoform and was also significantly increased in its mRNA level in HC islets [60]. These findings also indicate that the Ca^{2+}-independent augmentation pathway during the late phase of insulin secretion by HC islets is upregulated and that the glucose- and Ca^{2+}-independent pathway in HC islets is also activated by PKA and PKC (Fig. **4**). Although norepinephrine was found to completely inhibit insulin secretion by both MF and HC islets, an approximately 10-fold increase in norepinephrine in a dose-response curve was observed to achieve the same degree of inhibition in HC islets compared with MF islets, indicating reduced sensitivity of HC islets to norepinephrine [41].

Other major contributing factors to the development of obesity in HC rats are the altered energy circuitry mechanism of the hypothalamus and autonomic nervous system. Using specific agonists and antagonists of the parasympathetic and sympathetic nervous system, a significant increase in GSIS in response to cholinergic stimulation *via* parasympathetic input (Fig. **3**) as well as reduced sensitivity to adrenergic receptor-induced inhibition of insulin secretion *via* the sympathetic nervous system was observed in HC pups. Acetylcholine, a cholinergic agonist, enhanced to a greater extent GSIS by HC rats *in vivo* and from isolated islets, whereas sensitivity to inhibition of GSIS by oxymetazoline, an α_{2a}-adrenergic receptor ($\alpha_{2a}AR$) agonist, was reduced [61]. Furthermore, an increase in the mRNA levels of the cholinergic signaling molecules such as muscarinic type 3 receptor, phospholipase C-β1 and PKCα, and reduction in adrenergic receptor (α_{2a}AR) mRNA levels in islets from HC pups compared to MF pups lends support for an altered autonomic control of insulin secretion due to dietary changes [61]. These adaptations may play an important role in programming of hyperinsulinemia in response to the HC dietary intervention in the immediate postnatal period and the maintenance of postweaning hyperinsulinemia.

Although the growth rate of HC pups during the dietary treatment was not significantly different from that of age-matched MF pups, there were alterations in expression of orexigenic and anorexigenic neuropeptides in the hypothalamus of HC pups (Table **2**) [62]. The mRNA levels of neuropeptide Y and agouti-related protein were significantly increased whereas the mRNA levels of pro-opiomelanocortin, cocaine- and amphetamine-regulated transcript, corticotrophin-releasing factor and melanocortin-4-receptor were significantly decreased in the hypothalamus of HC pups, indicating programming effects for predisposal to hyperphagia in the immediate postweaning period.

Long-Term Consequences Due to Programming

Interestingly, despite weaning of HC rats on rodent laboratory chow on postnatal day 24, HC rats maintained hyperinsulinamia and experienced a hyperphagic response in the immediate postweaning period, resulting in increased weight gains by day 55 and developing obesity by day 100 [27, 30]. All programmed alterations at the cellular, molecular and biochemical levels in the islets of HC pups were also present in the islets from 100 day-old HC rats. These alterations included: (i) increased insulin producing mass [28], (ii) leftward shift in GSIS with increased secretory capacity [63], (iii) moderate amounts of insulin secretion in the absence of glucose and/or Ca^{2+} [63], (iv) increased activities of both low Km hexokinase and high Km glucokinase, (v) enhanced insulin secretion *via* a glucose- and Ca^{2+}-independent mechanism, (vi) altered autonomic nervous system regulation of insulin secretion (with increased parasympathetic input and reduced sympathetic tone) [64], (vii) increased expression of the *Pdx-1* and preproinsulin genes and (viii) significant increases in the expression of several cluster genes in islets from HC rats [28, 33, 59, 62, 63]. Furthermore, parasympathetic involvement in the maintenance of chronic hyperinsulinemia in adult HC rats is supported by a greater reduction in GSIS by vagotomy in HC rats compared with HC (intact) and MF (intact) rats [64].

Additionally, alterations in the gene expression of orexigenic and anorexigenic neuropeptides in the adult HC hypothalamus were similar to those observed in HC pups, indicating the programming effects of early dietary intervention in HC rats [62]. Altered methylation of specific CpG dinucleotides in the proximal promoter region of the *Npy* gene and increased acetylation of lysine 9 in histone 3 (H3K9) for the *Npy* gene were observed in hypothalami from adult HC rats compared with MF rats, supporting a possible mechanism explaining persistent hyperphagia [65]. Collectively, these findings clearly show that developmental plasticity initiated during the immediate postnatal period in response to nutritional intervention in HC rat islets continues into adulthood, resulting in the maintenance of hyperinsulinemia in the absence of any further nutritional challenge beyond day 24. However, the physiological consequence of increased food intake represents its own natural nutritional challenge in adult HC rats, which probably contributes to persistent hyperinsulinemia.

IS IT POSSIBLE TO REVERSE METABOLIC PROGRAMMING IN ADULTHOOD?

As discussed above, HC rats maintained hyperinsulinemia and also developed hyperphagia in the immediate post-weaning period, resulting in adult on-set obesity in these rats. Hence, two questions were tested for possible reversal of programming of β cells in adult HC rats. First, is it possible to normalize the HC phenotype (such as higher body weight gains and modified serum profiles of insulin and leptin) by caloric restriction during the postweaning period? Second, is it possible to permanently erase the phenotype of β cells in HC rats by caloric restriction? Using a pair-feeding procedure, daily food intake of HC rats was reduced to that of age-matched MF rats starting from weaning on postnatal day 24 and continued as indicated below. Pair-feeding of HC rats (referred to as HC/PF rats) up to postnatal day 140 amounted to reduction in average daily food intake over the entire restricted period to about 15% without causing any malnutrition. In answer to question 1, pair-feeding normalized postweaning body weight gains over the treatment period and normalized serum hormonal profiles (levels of insulin and leptin) in HC/PF rats to that of age-matched control MF rats [66]. These results indicated that it is possible to normalize the gross phenotype (such as body weight gains, serum insulin, and leptin levels) of HC rats by reducing their daily caloric intake to the level of control MF rats (question 1). To answer question 2, HC rats were pair-fed from the postnatal day 24 to day 90 and then switched to *ad libitum* feeding up to day 140 (referred to as HC/PF/AL rats). This treatment restored the HC metabolic phenotype (increased body weight gains and increased levels of serum insulin and leptin levels) in HC/PF/AL rats. Interestingly, the heightened GSIS of isolated islets from HC/PF and HC/PF/AL to glucose, acetylcholine (testing for the parasympathetic response), and oxymetazoline (the sympathetic response) were not significantly different from the responses of isolated islets from HC rats [66]. However, the hypothalamus was impacted less by the pair-feeding protocol than was the animal phenotype. The levels of *Npy* mRNA (orexigenic neuropeptide response) and pro-opiomelanocortin, cocaine (anorexigenic neuropeptide response) in the hypothalamus were not significantly different among HC, HC/PF, and HC/PF/AL rats on the postnatal day 140. Although the expression of the leptin receptor long form mirrored the levels of serum leptin levels in these three groups of rats, the levels of suppressor of cytokine signaling 3 mRNA remained elevated in these three groups, indicating that pair-feeding had no beneficial effect on reversal of programming in the HC hypothalamus. In sum, calorie restriction imposed by a pair-feeding procedure resulted in normalization of gross phenotypes of HC rats, but it was not efficacious for erasing the programmed effects in the β cells and the hypothalamus [66].

RELEVANCE TO OBESITY AND TYPE 2 DIABETES

During early stages of development, mammals experience time-dependent specific modifications in the composition of available nutrients. During fetal life, maternal circulation *via* the placenta provides a continuous supply of all nutrients with glucose being the primary source of energy. This mode of nutrient supply changes abruptly after birth when suckling is initiated by intermittent feeding of maternal milk (or equivalent source) enriched in fat or carbohydrate or both depending on the species. Towards the end of the suckling period, there again is a change in nutrient supply from milk-based to a solid diet for most mammals. During this fetal-newborn-weaning transition, many organs including the endocrine pancreas undergo structural and functional development. Any modifications in nutrient supply (either in quantity or quality or both) can have immediate as well as long-term consequences as has been amply documented under the original term as 'Fetal Origins of Adult Disease' and now with a new term as 'Developmental Origins of Health and Disease'. In westernized societies and especially in the United States, the practice of formula feeding (aka bottle feeding) with early introduction of carbohydrate-rich supplemental infant foods has resulted in alterations in both total calorie intake and nutrient composition of the mixed diet with increasing carbohydrate-derived calories. A modestly pre-matured weaning of rat pups (on postnatal day 18 instead of day 24) on a chow diet modified maturation of β cells by elevating the glucose-mediated mitogenic and secretory responses [67]. As summarized in this chapter, increased consumption of carbohydrate-derived calories during the suckling period (which may be considered as a form of precocious weaning on a diet high in carbohydrate-derived calories) markedly modified the islet structure and β cell function in rat pups. Extrapolation of these findings from rodents to humans raises a question about a link between the recent obesity epidemic in humans in general and childhood obesity in particular. Moreover, the precipitous decrease in breast-feeding and increased infant formula feeding together with early introduction of infant supplemental foods enriched in carbohydrate-derived calories are envisioned to contribute to changes in obesity prevalence in society. It is possible that even a modest increase in insulin secretory capacity of the β cell can have an accumulative effect on anabolic processes over a longer time period predisposing to the development of obesity and possibly to type 2 diabetes in adulthood.

CONFLICT OF INTEREST

The authors confirm that they have no conflict of interest to declare for this publication.

ACKNOWLEDGEMENTS

Research performed in the investigator's (MSP) laboratory and summarized in this chapter has been supported by National Institutes of Health Grants HD11089, DK61518 and DK51601.

REFERENCES

[1]   Barker DJ. The fetal and infant origins of disease. Eur J Clin Invest 1995; 25(7): 457-63.
        [http://dx.doi.org/10.1111/j.1365-2362.1995.tb01730.x] [PMID: 7556362]

[2]   McMillen IC, Adam CL, Mühlhäusler BS. Early origins of obesity: programming the appetite regulatory system. J Physiol 2005; 565(Pt 1): 9-17.
        [http://dx.doi.org/10.1113/jphysiol.2004.081992] [PMID: 15705647]

[3]   Fernandez-Twinn DS, Ozanne SE. Mechanisms by which poor early growth programs type-2 diabetes, obesity and the metabolic syndrome. Physiol Behav 2006; 88(3): 234-43.
        [http://dx.doi.org/10.1016/j.physbeh.2006.05.039] [PMID: 16782139]

[4]   Neu J, Hauser N, Douglas-Escobar M. Postnatal nutrition and adult health programming. Semin Fetal Neonatal Med 2007; 12(1): 78-86.
        [http://dx.doi.org/10.1016/j.siny.2006.10.009] [PMID: 17157087]

[5]   Rinaudo P, Wang E. Fetal programming and metabolic syndrome. Annu Rev Physiol 2012; 74: 107-30.
        [http://dx.doi.org/10.1146/annurev-physiol-020911-153245] [PMID: 21910625]

[6]   Nielsen JH, Haase TN, Jaksch C, *et al.* Impact of fetal and neonatal environment on β cell function and development of diabetes. Acta Obstet Gynecol Scand 2014; 93(11): 1109-22.
        [http://dx.doi.org/10.1111/aogs.12504] [PMID: 25225114]

[7]   Langley-Evans SC. Nutrition in early life and the programming of adult disease: a review. J Hum Nutr Diet 2015; 28 (Suppl. 1): 1-14.
        [http://dx.doi.org/10.1111/jhn.12212] [PMID: 24479490]

[8]   Kaung HL. Growth dynamics of pancreatic islet cell populations during fetal and neonatal development of the rat. Dev Dyn 1994; 200(2): 163-75.
        [http://dx.doi.org/10.1002/aja.1002000208] [PMID: 7919502]

[9]   Bonner-Weir S. β-cell turnover: its assessment and implications. Diabetes 2001; 50 (Suppl. 1): S20-4.
        [http://dx.doi.org/10.2337/diabetes.50.2007.S20] [PMID: 11272192]

[10]   Petrik J, Reusens B, Arany E, *et al.* A low protein diet alters the balance of islet cell replication and apoptosis in the fetal and neonatal rat and is associated with a reduced pancreatic expression of insulin-like growth factor-II. Endocrinology 1999; 140(10): 4861-73.
        [PMID: 10499546]

[11]   López-Soldado I, Munilla MA, Herrera E. Long-term consequences of under-nutrition during suckling on glucose tolerance and lipoprotein profile in female and male rats. Br J Nutr 2006; 96(6): 1030-7.
        [http://dx.doi.org/10.1017/BJN20061949] [PMID: 17181877]

[12]   Warner MJ, Ozanne SE. Mechanisms involved in the developmental programming of adulthood disease. Biochem J 2010; 427(3): 333-47.
        [http://dx.doi.org/10.1042/BJ20091861] [PMID: 20388123]

[13]   Godfrey KM, Gluckman PD, Hanson MA. Developmental origins of metabolic disease: life course and intergenerational perspectives. Trends Endocrinol Metab 2010; 21(4): 199-205.
[http://dx.doi.org/10.1016/j.tem.2009.12.008] [PMID: 20080045]

[14]   Alfaradhi MZ, Ozanne SE. Developmental programming in response to maternal overnutrition. Front Genet 2011; 2: 27.
[http://dx.doi.org/10.3389/fgene.2011.00027] [PMID: 22303323]

[15]   Barker DJ. Fetal origins of coronary heart disease. BMJ 1995; 311(6998): 171-4.
[http://dx.doi.org/10.1136/bmj.311.6998.171] [PMID: 7613432]

[16]   Neville MC, Anderson SM, McManaman JL, *et al.* Lactation and neonatal nutrition: defining and refining the critical questions. J Mammary Gland Biol Neoplasia 2012; 17(2): 167-88.
[http://dx.doi.org/10.1007/s10911-012-9261-5] [PMID: 22752723]

[17]   Woo JG, Martin LJ. Does breastfeeding protect against childhood obesity? Moving beyond observational evidence. Curr Obes Rep 2015; 4(2): 207-16.
[http://dx.doi.org/10.1007/s13679-015-0148-9] [PMID: 26627216]

[18]   James DC, Lessen R. Position of the American Dietetic Association: promoting and supporting breastfeeding. J Am Diet Assoc 2009; 109(11): 1926-42.
[http://dx.doi.org/10.1016/j.jada.2009.09.018] [PMID: 19862847]

[19]   Davidowa H, Plagemann A. Action of prolactin, prolactin-releasing peptide and orexins on hypothalamic neurons of adult, early postnatally overfed rats. Neuroendocrinol Lett 2005; 26(5): 453-8.
[PMID: 16264391]

[20]   Davidowa H, Plagemann A. Insulin resistance of hypothalamic arcuate neurons in neonatally overfed rats. Neuroreport 2007; 18(5): 521-4.
[http://dx.doi.org/10.1097/WNR.0b013e32805dfb93] [PMID: 17496815]

[21]   Rodrigues AL, De Souza EP, Da Silva SV, *et al.* Low expression of insulin signaling molecules impairs glucose uptake in adipocytes after early overnutrition. J Endocrinol 2007; 195(3): 485-94.
[http://dx.doi.org/10.1677/JOE-07-0046] [PMID: 18000310]

[22]   Liu HW, Mahmood S, Srinivasan M, Smiraglia DJ, Patel MS. Developmental programming in skeletal muscle in response to overnourishment in the immediate postnatal life in rats. J Nutr Biochem 2013; 24(11): 1859-69.
[http://dx.doi.org/10.1016/j.jnutbio.2013.05.002] [PMID: 23968580]

[23]   Srinivasan M, Laychock SG, Hill DJ, Patel MS. Neonatal nutrition: metabolic programming of pancreatic islets and obesity. Exp Biol Med (Maywood) 2003; 228(1): 15-23.
[PMID: 12524468]

[24]   Srinivasan M, Patel MS. Metabolic programming in the immediate postnatal period. Trends Endocrinol Metab 2008; 19(4): 146-52.
[http://dx.doi.org/10.1016/j.tem.2007.12.001] [PMID: 18329279]

[25]   Patel MS, Srinivasan M. Metabolic programming in the immediate postnatal life. Ann Nutr Metab 2011; 58 (Suppl. 2): 18-28.
[http://dx.doi.org/10.1159/000328040] [PMID: 21846978]

[26]   Haney PM, Estrin CR, Caliendo A, Patel MS. Precocious induction of hepatic glucokinase and malic enzyme in artificially reared rat pups fed a high-carbohydrate diet. Arch Biochem Biophys 1986; 244(2): 787-94.
[http://dx.doi.org/10.1016/0003-9861(86)90647-8] [PMID: 3511849]

[27]   Hiremagalur BK, Vadlamudi S, Johanning GL, Patel MS. Long-term effects of feeding high carbohydrate diet in pre-weaning period by gastrostomy: a new rat model for obesity. Int J Obes Relat Metab Disord 1993; 17(9): 495-502.
[PMID: 8220651]

[28]   Vadlamudi S, Hiremagalur BK, Tao L, *et al.* Long-term effects on pancreatic function of feeding a HC formula to rats during the preweaning period. Am J Physiol 1993; 265(4 Pt 1): E565-71.
[PMID: 8238331]

[29]   Patel MS, Vadlamudi S, Johanning GL. Artificial rearing of rat pups: implications for nutrition research. Annu Rev Nutr 1994; 14: 21-40.
[http://dx.doi.org/10.1146/annurev.nu.14.070194.000321] [PMID: 7946518]

[30]   Vadlamudi S, Kalhan SC, Patel MS. Persistence of metabolic consequences in the progeny of rats fed a HC formula in their early postnatal life. Am J Physiol 1995; 269(4 Pt 1): E731-8.
[PMID: 7485488]

[31]   Petrik J, Srinivasan M, Aalinkeel R, *et al.* A long-term high-carbohydrate diet causes an altered ontogeny of pancreatic islets of Langerhans in the neonatal rat. Pediatr Res 2001; 49(1): 84-92.
[http://dx.doi.org/10.1203/00006450-200101000-00019] [PMID: 11134497]

[32]   Srinivasan M, Song F, Aalinkeel R, Patel MS. Molecular adaptations in islets from neonatal rats reared artificially on a high carbohydrate milk formula. J Nutr Biochem 2001; 12(10): 575-84.
[http://dx.doi.org/10.1016/S0955-2863(01)00176-0] [PMID: 12031263]

[33]   Song F, Srinivasan M, Aalinkeel R, Patel MS. Use of a cDNA array for the identification of genes induced in islets of suckling rats by a high-carbohydrate nutritional intervention. Diabetes 2001; 50(9): 2053-60.
[http://dx.doi.org/10.2337/diabetes.50.9.2053] [PMID: 11522671]

[34]   Sharma A, Stein R. Glucose-induced transcription of the insulin gene is mediated by factors required for β-cell-type-specific expression. Mol Cell Biol 1994; 14(2): 871-9.
[http://dx.doi.org/10.1128/MCB.14.2.871] [PMID: 8289826]

[35]   Evans-Molina C, Garmey JC, Ketchum R, Brayman KL, Deng S, Mirmira RG. Glucose regulation of insulin gene transcription and pre-mRNA processing in human islets. Diabetes 2007; 56(3): 827-35.
[http://dx.doi.org/10.2337/db06-1440] [PMID: 17327454]

[36]   Poitout V, Hagman D, Stein R, Artner I, Robertson RP, Harmon JS. Regulation of the insulin gene by glucose and fatty acids. J Nutr 2006; 136(4): 873-6.
[PMID: 16549443]

[37]   Rafiq I, Kennedy HJ, Rutter GA. Glucose-dependent translocation of insulin promoter factor-1 (IPF-1) between the nuclear periphery and the nucleoplasm of single MIN6 β-cells. J Biol Chem 1998; 273(36): 23241-7.
[http://dx.doi.org/10.1074/jbc.273.36.23241] [PMID: 9722555]

[38]   Macfarlane WM, McKinnon CM, Felton-Edkins ZA, Cragg H, James RF, Docherty K. Glucose stimulates translocation of the homeodomain transcription factor PDX1 from the cytoplasm to the nucleus in pancreatic β-cells. J Biol Chem 1999; 274(2): 1011-6.
[http://dx.doi.org/10.1074/jbc.274.2.1011] [PMID: 9873045]

[39]   Porat S, Weinberg-Corem N, Tornovsky-Babaey S, *et al.* Control of pancreatic β cell regeneration by glucose metabolism. Cell Metab 2011; 13(4): 440-9.
[http://dx.doi.org/10.1016/j.cmet.2011.02.012] [PMID: 21459328]

[40]   Patel MS, Srinivasan M, Strutt B, Mahmood S, Hill DJ. Featured Article: β cell specific pyruvate dehydrogenase alpha gene deletion results in a reduced islet number and β-cell mass postnatally. Exp Biol Med (Maywood) 2014; 239(8): 975-85.
[http://dx.doi.org/10.1177/1535370214531895] [PMID: 24845368]

[41]   Aalinkeel R, Srinivasan M, Kalhan SC, Laychock SG, Patel MS. A dietary intervention (high carbohydrate) during the neonatal period causes islet dysfunction in rats. Am J Physiol 1999; 277(6 Pt 1): E1061-9.
[PMID: 10600796]

[42]   Withers DJ, Gutierrez JS, Towery H, *et al.* Disruption of IRS-2 causes type 2 diabetes in mice. Nature 1998; 391(6670): 900-4.
[http://dx.doi.org/10.1038/36116] [PMID: 9495343]

[43]   Kubota N, Tobe K, Terauchi Y, *et al.* Disruption of insulin receptor substrate 2 causes type 2 diabetes because of liver insulin resistance and lack of compensatory β-cell hyperplasia. Diabetes 2000; 49(11): 1880-9.
[http://dx.doi.org/10.2337/diabetes.49.11.1880] [PMID: 11078455]

[44]   Lee YC, Nielsen JH. Regulation of β cell replication. Mol Cell Endocrinol 2009; 297(1-2): 18-27.
[http://dx.doi.org/10.1016/j.mce.2008.08.033] [PMID: 18824066]

[45]   Bernardo AS, Hay CW, Docherty K. Pancreatic transcription factors and their role in the birth, life and survival of the pancreatic β cell. Mol Cell Endocrinol 2008; 294(1-2): 1-9.
[http://dx.doi.org/10.1016/j.mce.2008.07.006] [PMID: 18687378]

[46]   Kitamura T, Nakae J, Kitamura Y, *et al.* The forkhead transcription factor Foxo1 links insulin signaling to Pdx1 regulation of pancreatic β cell growth. J Clin Invest 2002; 110(12): 1839-47.
[http://dx.doi.org/10.1172/JCI200216857] [PMID: 12488434]

[47]   Kaneto H, Matsuoka TA. Role of pancreatic transcription factors in maintenance of mature β-cell function. Int J Mol Sci 2015; 16(3): 6281-97.
[http://dx.doi.org/10.3390/ijms16036281] [PMID: 25794287]

[48]   Patel MS, Srinivasan M. Metabolic programming due to alterations in nutrition in the immediate postnatal period. J Nutr 2010; 140(3): 658-61.
[http://dx.doi.org/10.3945/jn.109.110155] [PMID: 20107149]

[49]   Detimary P, Gilon P, Henquin JC. Interplay between cytoplasmic Ca^{2+} and the ATP/ADP ratio: a feedback control mechanism in mouse pancreatic islets. Biochem J 1998; 333(Pt 2): 269-74.
[http://dx.doi.org/10.1042/bj3330269] [PMID: 9657965]

[50]   Schuit FC, Huypens P, Heimberg H, Pipeleers DG. Glucose sensing in pancreatic β-cells: a model for

the study of other glucose-regulated cells in gut, pancreas, and hypothalamus. Diabetes 2001; 50(1): 1-11.
[http://dx.doi.org/10.2337/diabetes.50.1.1] [PMID: 11147773]

[51]　Roduit R, Nolan C, Alarcon C, *et al*. A role for the malonyl-CoA/long-chain acyl-CoA pathway of lipid signaling in the regulation of insulin secretion in response to both fuel and nonfuel stimuli. Diabetes 2004; 53(4): 1007-19.
[http://dx.doi.org/10.2337/diabetes.53.4.1007] [PMID: 15047616]

[52]　MacDonald MJ, Fahien LA, Brown LJ, Hasan NM, Buss JD, Kendrick MA. Perspective: emerging evidence for signaling roles of mitochondrial anaplerotic products in insulin secretion. Am J Physiol Endocrinol Metab 2005; 288(1): E1-E15.
[http://dx.doi.org/10.1152/ajpendo.00218.2004] [PMID: 15585595]

[53]　Jensen MV, Joseph JW, Ilkayeva O, *et al*. Compensatory responses to pyruvate carboxylase suppression in islet β-cells. Preservation of glucose-stimulated insulin secretion. J Biol Chem 2006; 281(31): 22342-51.
[http://dx.doi.org/10.1074/jbc.M604350200] [PMID: 16740637]

[54]　Guay C, Madiraju SR, Aumais A, Joly E, Prentki M. A role for ATP-citrate lyase, malic enzyme, and pyruvate/citrate cycling in glucose-induced insulin secretion. J Biol Chem 2007; 282(49): 35657-65.
[http://dx.doi.org/10.1074/jbc.M707294200] [PMID: 17928289]

[55]　Rorsman P, Arkhammar P, Bokvist K, *et al*. Failure of glucose to elicit a normal secretory response in fetal pancreatic β cells results from glucose insensitivity of the ATP-regulated K^+ channels. Proc Natl Acad Sci USA 1989; 86(12): 4505-9.
[http://dx.doi.org/10.1073/pnas.86.12.4505] [PMID: 2543980]

[56]　Gu C, Stein GH, Pan N, *et al*. Pancreatic β cells require NeuroD to achieve and maintain functional maturity. Cell Metab 2010; 11(4): 298-310.
[http://dx.doi.org/10.1016/j.cmet.2010.03.006] [PMID: 20374962]

[57]　Aguayo-Mazzucato C, Koh A, El Khattabi I, *et al*. Mafa expression enhances glucose-responsive insulin secretion in neonatal rat β cells. Diabetologia 2011; 54(3): 583-93.
[http://dx.doi.org/10.1007/s00125-010-2026-z] [PMID: 21190012]

[58]　Guo L, Inada A, Aguayo-Mazzucato C, *et al*. PDX1 in ducts is not required for postnatal formation of β-cells but is necessary for their subsequent maturation. Diabetes 2013; 62(10): 3459-68.
[http://dx.doi.org/10.2337/db12-1833] [PMID: 23775765]

[59]　Laychock SG, Vadlamudi S, Patel MS. Neonatal rat dietary carbohydrate affects pancreatic islet insulin secretion in adults and progeny. Am J Physiol 1995; 269(4 Pt 1): E739-44.
[PMID: 7485489]

[60]　Srinivasan M, Aalinkeel R, Song F, Lee B, Laychock SG, Patel MS. Adaptive changes in insulin secretion by islets from neonatal rats raised on a high-carbohydrate formula. Am J Physiol Endocrinol Metab 2000; 279(6): E1347-57.
[PMID: 11093923]

[61]　Mitrani P, Srinivasan M, Dodds C, Patel MS. Role of the autonomic nervous system in the development of hyperinsulinemia by high-carbohydrate formula feeding to neonatal rats. Am J Physiol Endocrinol Metab 2007; 292(4): E1069-78.

[http://dx.doi.org/10.1152/ajpendo.00477.2006] [PMID: 17164433]

[62]   Srinivasan M, Mitrani P, Sadhanandan G, *et al.* A high-carbohydrate diet in the immediate postnatal life of rats induces adaptations predisposing to adult-onset obesity. J Endocrinol 2008; 197(3): 565-74.
[http://dx.doi.org/10.1677/JOE-08-0021] [PMID: 18492820]

[63]   Aalinkeel R, Srinivasan M, Song F, Patel MS. Programming into adulthood of islet adaptations induced by early nutritional intervention in the rat. Am J Physiol Endocrinol Metab 2001; 281(3): E640-8.
[PMID: 11500321]

[64]   Mitrani P, Srinivasan M, Dodds C, Patel MS. Autonomic involvement in the permanent metabolic programming of hyperinsulinemia in the high-carbohydrate rat model. Am J Physiol Endocrinol Metab 2007; 292(5): E1364-77.
[http://dx.doi.org/10.1152/ajpendo.00672.2006] [PMID: 17227957]

[65]   Mahmood S, Smiraglia DJ, Srinivasan M, Patel MS. Epigenetic changes in hypothalamic appetite regulatory genes may underlie the developmental programming for obesity in rat neonates subjected to a high-carbohydrate dietary modification. J Dev Orig Health Dis 2013; 4(6): 479-90.
[http://dx.doi.org/10.1017/S2040174413000238] [PMID: 24924227]

[66]   Srinivasan M, Mahmood S, Patel MS. Metabolic programming effects initiated in the suckling period predisposing for adult-onset obesity cannot be reversed by calorie restriction. Am J Physiol Endocrinol Metab 2013; 304(5): E486-94.
[http://dx.doi.org/10.1152/ajpendo.00519.2012] [PMID: 23249696]

[67]   Stolovich-Rain M, Enk J, Vikesa J, *et al.* Weaning triggers a maturation step of pancreatic β cells. Dev Cell 2015; 32(5): 535-45.
[http://dx.doi.org/10.1016/j.devcel.2015.01.002] [PMID: 25662175]

CHAPTER 6

New Concepts in the Intra-Islet Control of β-Cell Function and Mass

Brian T. Layden[1,2,*], Stephanie Villa[1] and **William L. Lowe[1]**

[1] *Division of Endocrinology, Metabolism and Molecular Medicine, Northwestern University Feinberg School of Medicine, Chicago, Illinois, USA*

[2] *Jesse Brown Veterans Affairs Medical Center, Chicago, Illinois, USA*

Abstract: The regulation of pancreatic β-cell function and mass is critical to the maintenance of euglycemia. β-cells integrate numerous signals from the host to secrete appropriate amounts of insulin and maintain tight control of blood glucose levels. Together with glucose; nutrients, amino acids, hormones, and metabolic by-products contribute to this physiologic response. Within the islet microenvironment, where β-cells reside, there exists a network of interacting pathways that contribute to insulin secretion and regulation of β-cell mass. While factors within these pathways are often sourced from digestive processes and peripheral tissues, intra-islet-derived factors are also important components in the ability of -βcells to accurately integrate metabolic demands with β-cell function. In recent years, many biologic factors have been found to have previously unappreciated autocrine and paracrine roles within the islet. Moreover, differences have been described between signaling within rodent and human islets that are important for informing our understanding of autocrine/paracrine signaling between species. In this review, we highlight these new findings and future directions for this field of study.

Keywords: Autocrine, β-cells, β-cell function, β-cell mass, Diabetes, Glucose, Insulin, Islets, Nutrient-sensing, Paracrine.

* **Corresponding author Brian T. Layden:** Division of Endocrinology, Metabolism, and Molecular Medicine, Department of Medicine, Northwestern University, Chicago, IL, USA; Tel/Fax: 312-503-0006/312-908-9032; E-mail: b-layden@northwestern.edu

David J. Hill (Ed.)

INTRODUCTION

Defining the Importance of β-Cell Function and Mass

The Islets of Langerhans are composed of multiple cell types. β-cells, the most abundant cell type within islets, produce and secrete insulin in response to glucose. The metabolism of glucose initiates biochemical signaling cascades that ultimately result in β-cell depolarization, insulin granule fusion, and insulin secretion. The secretion of insulin, which is synthesized primarily in pancreatic islets, is largely dependent on extracellular glucose levels and tightly regulated to maintain euglycemia. While glucose is the primary mediator of insulin secretion, other factors (such as nutrients, cytokines, and hormones) fine-tune insulin secretion based on the physiologic needs of the organism. Deficient insulin secretion is central to the pathogenesis of diabetes mellitus (DM) and results in diminished uptake of glucose by peripheral tissues, which leads to hyperglycemia. Understanding the additional pathways involved in the regulation of insulin secretion and maintenance of glucose homeostasis has provided novel mechanistic insight into DM and pathways to target for its treatment.

While blood glucose levels are the primary driver of insulin secretion, changes in insulin sensitivity that accompany physiologic conditions such as pregnancy [1] and obesity [2] require the pancreas to increase its capacity for insulin secretion. To support this demand, β-cells have the capacity to expand their overall mass through the regulation of β-cell proliferation and apoptosis, β-cell hypertrophy, and/or β-cell neogenesis [3]. However, the pathways and mechanisms (proliferation/apoptosis, hypertrophy or neogenesis) that predominate are still debated. Regardless, the end result of β-cell mass expansion is to provide increased insulin secretory capacity under conditions of increased insulin demand. Because of this, identification of endogenous and exogenous factors that promote β-cell mass expansion provides insight into mechanisms used to compensate for deficient insulin secretion and its alteration in states such as type 2 diabetes (T2D) and gestational diabetes mellitus (GDM).

Known Factors Affecting β-Cell Function and Mass and Their Mode of Action

As noted above, glucose is the primary insulin secretagogue, and elevated glucose has been shown to increase β-cell mass by inducing β-cell proliferation in both mouse and human islets [4, 5]. Along with glucose, other nutrients (fatty acids and amino acids) and hormones such as glucagon-like peptide-1 (GLP-1) have a documented role in insulin secretion and β-cell mass regulation (Table 1). In general, these nutrients and hormones are derived either from dietary sources or by secretion from a peripheral tissue and delivered to islets through the systemic circulation. Thus, these factors act as endocrine hormones. However, hormones can also be secreted and signal locally through a paracrine (action on neighboring cells) or autocrine (action on cells the factor was secreted by) effect.

Table 1. Partial listing of extracellular signaling factors affecting insulin secretion and β-cell proliferation.

	Signal type	Mode of action at β-cell	β-cell effect
Insulin	Autocrine	Receptor tyrosine kinase	Potentiate GSIS?
GLP-1 [6]	Endocrine/Paracrine	GPCR	Potentiate GSIS Stimulate β-cell proliferation Protect against β-cell death
Glucose-dependent insulinotropic polypeptide (GIP) [6]	Endocrine	GPCR	Potentiate GSIS Stimulate β-cell proliferation Protect against β-cell death
Somatostatin [7]	Paracrine	GPCR	Inhibit GSIS
Amino acids [8]	Endocrine	GPCR/ Intracellular metabolism	Inhibit or potentiate GSIS (depending on amino acid type)
Ghrelin [7]	Endocrine/Paracrine/Autocrine	GPCR	Inhibit GSIS
Vasoactive intestinal polypeptide (VIP) [7]	Neuropeptide	GPCR	Potentiate GSIS

(Table 1) contd.....

	Signal type	Mode of action at β-cell	β-cell effect
Pituitary adenylate cyclase activating polypeptide (PACAP) [7]	Neuropeptide	GPCR	Potentiate GSIS
Cholecystokinin (CCK) [6]	Endocrine	GPCR	Potentiate GSIS Stimulate β-cell proliferation
GABA [9]	Paracrine	Ionotropic receptors, GPCRs	Potentiate GSIS Promote β-cell survival
Acetylcholine [10, 11]	Paracrine	GPCR	Potentiate GSIS
Islet amyloid polypeptide (IAPP) [12]	Autocrine	GPCR	Inhibit GSIS Promote β-cell proliferation
Fatty Acids [8]	Endocrine	GPCR/ Intracellular metabolism	Potentiate or inhibit GSIS (depending on fatty acid type) May increase β-cell mass

Each nutrient and/or hormone acts through unique modes of action in their regulation of β-cell function and mass, ranging from receptor-dependent (including G protein-coupled receptor (GPCR)) signaling to receptor-independent pathways such as anaplerosis, a series of biochemical reactions that replenishes components of the citric acid cycle [13] (Fig. **1**). Moreover, the mode of action can vary, depending on whether these hormones are affecting insulin secretion or β-cell mass. However, significant redundancy exists in the signaling mechanisms by which hormones exert their mode of action. The goal of this review is to focus on newly identified islet-derived factors (either paracrine or autocrine) that have been suggested to have a role in the regulation of β-cell function and mass.

PARACRINE AND AUTOCRINE FACTORS REGULATING β-CELL FUNCTION

Anatomic Differences Affect Paracrine and Autocrine Signaling Properties Between Species

When considering paracrine/autocrine factors and their role in β-cell function, the islet microenvironment is an important consideration, and how this environment differs between species is noteworthy. Through immunofluorescence studies, it has been well-recognized that islet topology is distinct between humans and rodents, which is an important distinction, considering that a large portion of published studies in this area of research use only rodent islets as a model of β-cell biology. Specifically, in humans, β-cells are dispersed throughout the islet, while in rodents, β-cells are concentrated in the central region and surrounded by α and other endocrine cells [14]. The percentage of β-cells is also much lower in humans compared to rodents (60% and 90%, respectively) [15]. It is anticipated that these differences may lead to differences in the function of paracrine/autocrine factors between species. For example, in human islets, paracrine interaction between β-cells and other islet-cell types is more feasible than in rodents [16]. Therefore, any data on paracrine/autocrine factors from rodents clearly cannot be directly extrapolated to human islets and need to be confirmed in human islet studies [16].

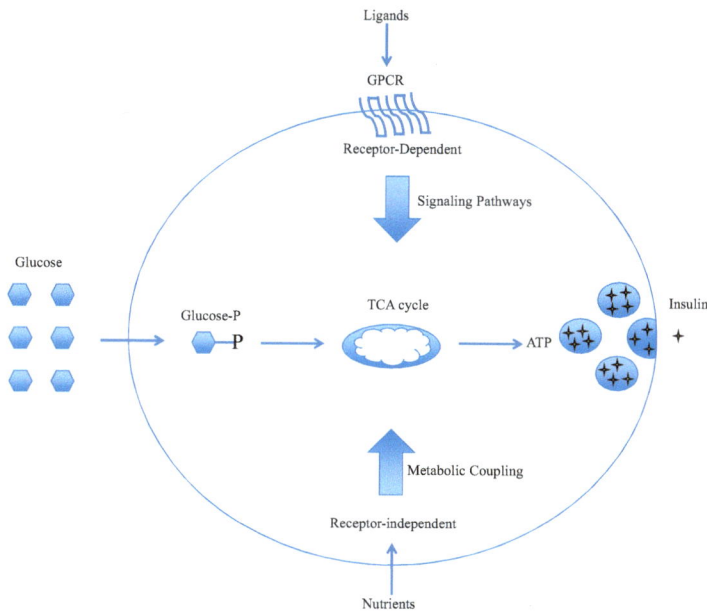

Fig. (1). Depiction of the mode of action for glucose and other common pathways (receptor-dependent or independent) involved in mediating insulin secretion. Receptor-dependent signaling commonly occurs *via* GPCRs, and the ligands for GPCRs range from nutrients to hormones. Contributions from receptor-independent signaling occur with metabolic coupling *via* factors derived from intracellular nutrient metabolism.

Another important variable that warrants consideration is the vasculature of islets, which is highly dense, and facilitates delivery of islet hormones into the systemic circulation. The direction of blood flow through the islet vasculature is an important determinant in the ability of paracrine factors to reach their target tissues. In mouse, it has been suggested that blood flow occurs primarily from the inside of the islet outward, *i.e.*, perfusion from the β-cell out. However, in a subset of islets, it was observed that capillaries demonstrated extensive branching immediately upon entering the islet, allowing perfusion of blood from one side of the islet to the other [17]. It is not well explored how the pattern of vascularization in the mouse compares to that of human islets.

Emerging Autocrine Factors

Insulin

Studies over the past few years from Goldfine and Kulkarni have demonstrated that insulin regulates its own secretion. These authors explored the role of insulin signaling in β-cells in both rodent and human islets based on the expression of insulin receptors in β-cells. In mice, β-cell-specific knockout of the insulin receptor impairs insulin secretion, and with aging, these mice develop impaired glucose tolerance and hyperglycemia [18]. Translating these observations to humans, pre-exposure to insulin enhances glucose stimulated insulin secretion in humans [19], and this enhancement is diminished in T2D [20]. While most classic hormones negatively regulate their own secretion, these data suggest that insulin may potentiate its own secretion.

However, whether this intriguing autocrine/paracrine-insulin loop is actually possible is a matter of some debate. Specifically, Rhodes, *et al.* have suggested that, exposure of islets to high levels of insulin *in vivo* would likely lead to downregulation of insulin receptors, thus limiting the potential for autocrine signaling to occur [21]. Alternatively, the authors argue that the intra-islet environment may contain very little insulin, as it is likely that insulin is efficiently shuttled into the general circulation, with concentrations greatly decreased by the time circulation is returned to the β-cell, owing to significant degradation and uptake by other tissues. Thus, these authors propose that little opportunity exists for autocrine signaling by insulin [21]. These authors suggest an alternative indirect mechanism by which insulin exerts its effects on β-cells *via* the central nervous system. If true, the role of the β-cell's extensive insulin receptor-signaling pathway remains uncertain. If insulin does not activate its cognate receptor in β-cells, another possibility is that ligands such as insulin-like growth factors (IGFs) may activate insulin-receptor signaling pathways [22]. Regardless, this unresolved controversy has direct clinical implications and warrants further exploration. Specifically, timing of insulin therapy initiation is often debated for T2D patients, and better understanding of the effect of exogenous insulin on β-cell function may impact the approach to treatment of T2D.

Proinsulin C-Peptide

The connecting peptide, or C-peptide, is a byproduct of the insulin synthesis pathway, resulting from cleavage of proinsulin to insulin [23]. Previously thought to be biologically inactive, accumulating data suggest that C-peptide is in fact bioactive, binding to cell membranes and inducing intracellular signaling cascades [24]. Although the mechanism by which C-peptide elicits these effects is not fully understand, evidence suggests that these effects may be, in part, mediated by a G protein-coupled receptor (GPCR), GPR146 [25]. Recently, C-peptide was shown to inhibit both glucose- and arginine-dependent insulin secretion in clonal β-cells and isolated mouse islets [26]. Independent confirmation is needed for these thought-provoking findings, as well as exploration of a role for C-peptide in human islets.

Islet Amyloid Polypeptide

β-cells co-secrete Islet Amyloid Polypeptide (IAPP) with insulin, and it has been suggested that IAPP accumulation, which leads to amyloid deposits in islets, may contribute to the pathogenesis of T2D. However, IAPP has also been shown to affect β-cell function, specifically by inhibiting glucose-stimulated insulin secretion. While many of these studies were done over 20 years ago, we discuss IAPP here because a recent report suggests that IAPP impacts β-cell mass *via* a previously unexplored autocrine mechanism of action. Specifically, IAPP acts as mitogenic factor in rodent β-cells in a glucose-dependent manner (with the mitogenic effect of IAPP diminishing in high glucose) [12]. It is known that IAPP acts *via* a GPCR-dependent mechanism through the calcitonin receptor and the associated receptor activity-modifying proteins (RAMP) 1 – 3. This raises the possibility that IAPP signaling *via* the calcitonin receptor may be a novel β-cell proliferative pathway and suggests the possible use of IAPP as a regenerative β-cell factor. However, careful consideration of IAPP as a therapeutic approach is necessary given its inhibitory effect on insulin secretion which could exacerbate hyperglycemia in the setting of diabetes.

GABA

A role for this neurotransmitter in influencing islet biology has been suggested for decades [27], and new data have recently emerged from both rodent and human islet studies. As noted above, human and mouse islet biology is distinct, and human and mouse islets seem to differ in their capacity to produce GABA. Human islet endocrine cells (α, β, and δ cells) all contain GABA [9]; whereas in mouse islets, GABA seems to be particularly concentrated in β-cells. In a relatively recent report, Braun *et al.* [9] showed that GABA was secreted by β-cells *via* glucose-dependent and independent mechanisms and that GABA augments insulin secretion through the GABA-A receptor in human islets [9]. These findings are consistent with GABA mediating its effects *via* a novel autocrine/paracrine feedback loop. In rodents, exogenous administration of GABA has a protective effect on the development of type 1 diabetes by preventing the development of insulitis and, thus, preserving β-cell mass [28].

Serotonin

The role of this well-described neurotransmitter in the β-cell emerged from pregnancy studies in rodents. Specifically, gene expression studies demonstrated that tryptophan hydroxylase, the rate-limiting enzyme in the synthesis of serotonin (5-HT), is dramatically increased in islets during pregnancy [29, 30] resulting in increased 5-HT within the islet environment. The consequence of elevated 5-HT during pregnancy is increased insulin secretion and β-cell proliferation, which is mediated by the 5-HT receptors HTR2b and HTR3, respectively [31, 32]. However, whether a similar 5-HT autocrine loop exists in human islets has not been reported yet, and further study is needed.

Nicotonic Acid Adenine Dinucleotide Phosphate (NAADP)

NAADP is a second messenger signaling molecule and a potent stimulator of intracellular calcium mobilization. In a recent report, NAADP was shown to be released from islets and the MIN6 β-cell line following glucose stimulation [33]; however, as the MIN6 cell line appears to be a mixed cell line [34], it is not conclusively known from what cell type NAADP is released. Nonetheless, this study revealed that extracellular application of NAADP led to potentiation of GSIS in islets, and exogenous NAADP administration improved glucose tolerance in diabetic mice *in vivo*, in part, as a result of improved insulin secretion [33]. However, it is currently unclear how NAADP exerts these effects, although the possibility of its signaling *via* GPCRs has been suggested.

Cholecystokinin (CCK)

CCK is a well-described peptide secreted from the small intestine in response to meals. Over 30 years ago, CCK was observed to contribute to insulin secretion [35]. More recently, it was discovered that CCK is expressed in β-cells in the setting of obesity, and that loss of its expression leads to diminished β-cell mass and increased apoptosis during obesity [36]. Exploring this further, it was found that secretion of glucagon-like peptide 1 (GLP-1) by islet α cells leads to CCK transcription and secretion in β-cells, and that part of the protective effect of GLP-1 on apoptosis is mediated by CCK signaling in islets [36, 37]. Taken together, these data suggest that paracrine signaling by GLP-1 between α and β-cells leads to β-cell induction of CCK, which may exert a protective effect on β-cells through autocrine signaling [37].

Acetate

It is well described that long and medium chain fatty acids impact β-cell function, but there has been less exploration into the role of short chain fatty acids (SCFAs), which include the molecules acetate, propionate, and butyrate [38]. A recent report by Tang, *et al.* suggested that glucose metabolism by β-cells leads to acetate production which results in autocrine inhibition of GSIS *via* two SCFAspecific GPCRs (free fatty acid receptor-2 and -3; FFAR2 and FFAR3) [39]. However, two other reports exploring the effect of FFAR2 signaling in β-cells have contradicted these findings, reporting that signaling *via* FFAR2 potentiates insulin secretion [40, 41]. These discrepancies extend to studies of insulin secretion by human islets with Tang, *et al.* reporting that acetate inhibits GSIS, while the other two studies found no direct effect of acetate on GSIS. Thus, it is clear that future studies will be necessary to fully understand the role of SCFAs in regulating β-cell function.

Emerging Paracrine Factors

Acetylcholine

Previously, most studies suggested that acetylcholine (ACh) modulates β-cell function *via* ACh secretion from parasympathetic neurons that innervate the islet. Important species differences between rodent and human islets exist here; while mouse islets are highly innervated, human islets are not innervated with cholinergic neurons [42]. However, human α cells secrete ACh and influence β-cell function by signaling through the M3 and M5 muscarinic receptor subtypes, leading to enhanced insulin secretion [10, 11]. ACh can also act at δ cells through the M1 muscarinic receptor, leading to somatostatin secretion. Thus, ACh can influence insulin secretion *via* direct signaling at the β-cell and through indirect mechanisms mediated by somatostatin [11]. These elegant studies highlight the importance of translating findings to human islets.

VGF-Derived Peptides

Vgf is a gene that encodes a precursor protein which, following proteolysis, produces an array of peptides; many of which are known to have metabolic effects [43]. Importantly, these peptides are expressed in β-cells and islets [44], and early studies have suggested a role in β-cell function [45, 46]. In the first published report, Stephens, *et al.* found that *Vgf* expression is strongly upregulated in response to overexpression of Nkx6.1, a homeobox transcription factor that enhances insulin secretion and β-cell replication [46, 47]. Further investigation revealed that the VGF-derived peptide, TLQP-21, which was previously reported to have a role in central nervous system-mediated control of metabolism [48], potentiates glucose stimulated insulin secretion and protects against β-cell apoptosis [46]. In another report, screening of multiple VGF-derived peptides indicated that TLQP-62 may be the most potent regulator of GSIS among VGF-derived peptides [45]. Future studies into the VGF family of peptides will be needed to identify the receptor(s) involved in VGF-peptide signaling. Likewise, the islet cell type from which these VGF-derived peptides are derived needs to be determined, as does how these peptides function within human islets. Stephens *et al.* observed that TLQP-21 is secreted in response to elevated glucose levels; however, it is not clear whether the origin of TLQP-21 is the β-cell or another cell type within the islet [46].

Approaches to Identify New Autocrine/Paracrine Signaling Factors

In a recent paper, Yang *et al.*, (2011) used multiple approaches to identify potential novel signaling pathways in mouse and human islets [49]. Using gene expression databases, islet-specific tag sequencing libraries, and microarray datasets from purified β-cells, the authors identified 190 ligand/receptor pairs from a total of 233 secreted factors and 234 receptors for secreted factors [49]. These data led the authors to identify a previously unidentified role of netrins and their receptors in β-cell apoptosis [49]. Using this resource, the authors next explored another novel pathway, Slit ligands and their receptors, the Roundabout receptors (Robo), and uncovered a role for these factors in β-cell survival and insulin secretion [50]. This database suggests the high level of complexity that may exist in the regulation of β-cell function and insulin secretion in islets.

CONCLUSION

The past few years have revealed a wealth of data on new autocrine/paracrine factors that contribute to β-cell function. While many of these recent reports seem to highlight important novel regulators of β-cell function and mass, differences between mouse and human islet topology and physiology make it critical that these experimental observations are verified in human islet studies. Additionally, as seen by the number of potential ligand/receptor pairs, a complete picture of the intra-islet signaling pathways that mediate insulin secretion is far from complete. Moving forward, it will be necessary to reconcile our understanding of these diverse intra-islet pathways and how they co-exist with pathways outside the micro-environment of the islet.

CONFLICT OF INTEREST

The authors confirm that they have no conflict of interest to declare for this publication.

ACKNOWLEDGEMENTS

BTL is supported by the Department of Veterans Affairs, Veterans Health Administration, Office of Research and Development, Career Development Grant #1IK2BX001587-01. S.R.V. is supported by the National Institute of Diabetes and Digestive and Kidney Diseases of the National Institutes of Health (F31DK102371), the Northwestern University Program in Endocrinology, Diabetes and Hormone Action (NIH T32 DK007169), and the Northwestern University Cellular and Molecular Basis of Disease training grant (NIH T32 GM08061).

REFERENCES

[1]   Butler AE, Cao-Minh L, Galasso R, *et al.* Adaptive changes in pancreatic β cell fractional area and β cell turnover in human pregnancy. Diabetologia 2010; 53(10): 2167-76.
[http://dx.doi.org/10.1007/s00125-010-1809-6] [PMID: 20523966]

[2]   Saisho Y, Butler AE, Manesso E, Elashoff D, Rizza RA, Butler PC. β-cell mass and turnover in humans: effects of obesity and aging. Diabetes Care 2013; 36(1): 111-7.
[http://dx.doi.org/10.2337/dc12-0421] [PMID: 22875233]

[3]   Jurczyk A, Bortell R, Alonso LC. Human β-cell regeneration: progress, hurdles, and controversy. Curr Opin Endocrinol Diabetes Obes 2014; 21(2): 102-8.
[http://dx.doi.org/10.1097/MED.0000000000000042] [PMID: 24569551]

[4]   Alonso LC, Yokoe T, Zhang P, *et al.* Glucose infusion in mice: a new model to induce β-cell replication. Diabetes 2007; 56(7): 1792-801.
[http://dx.doi.org/10.2337/db06-1513] [PMID: 17400928]

[5]   Levitt HE, Cyphert TJ, Pascoe JL, *et al.* Glucose stimulates human β cell replication *in vivo* in islets transplanted into NOD-severe combined immunodeficiency (SCID) mice. Diabetologia 2011; 54(3): 572-82.
[http://dx.doi.org/10.1007/s00125-010-1919-1] [PMID: 20936253]

[6]   Lavine JA, Attie AD. Gastrointestinal hormones and the regulation of β-cell mass. Ann N Y Acad Sci 2010; 1212: 41-58.
[http://dx.doi.org/10.1111/j.1749-6632.2010.05802.x] [PMID: 21039588]

[7]   Koh DS, Cho JH, Chen L. Paracrine interactions within islets of Langerhans. J Mol Neurosci 2012; 48(2): 429-40.
[http://dx.doi.org/10.1007/s12031-012-9752-2] [PMID: 22528452]

[8]   Newsholme P, Cruzat V, Arfuso F, Keane K. Nutrient regulation of insulin secretion and action. J Endocrinol 2014; 221(3): R105-20.
[http://dx.doi.org/10.1530/JOE-13-0616] [PMID: 24667247]

[9]   Braun M, Ramracheya R, Bengtsson M, *et al.* Gamma-aminobutyric acid (GABA) is an autocrine excitatory transmitter in human pancreatic β-cells . Diabetes 2010; 59(7): 1694-701.
[http://dx.doi.org/10.2337/db09-0797] [PMID: 20413510]

[10]   Rodriguez-Diaz R, Dando R, Jacques-Silva MC, *et al.* Alpha cells secrete acetylcholine as a non-neuronal paracrine signal priming β cell function in humans. Nat Med 2011; 17(7): 888-92.
[http://dx.doi.org/10.1038/nm.2371] [PMID: 21685896]

[11]   Molina J, Rodriguez-Diaz R, Fachado A, Jacques-Silva MC, Berggren PO, Caicedo A. Control of insulin secretion by cholinergic signaling in the human pancreatic islet. Diabetes 2014; 63(8): 2714-26.
[http://dx.doi.org/10.2337/db13-1371] [PMID: 24658304]

[12]   Visa M, Alcarraz-Vizán G, Montane J, *et al.* Islet amyloid polypeptide exerts a novel autocrine action in β-cell signaling and proliferation. FASEB J 2015; 29(7): 2970-9.
[http://dx.doi.org/10.1096/fj.15-270553] [PMID: 25808537]

[13]   Prentki M, Matschinsky FM, Madiraju SR. Metabolic signaling in fuel-induced insulin secretion. Cell Metab 2013; 18(2): 162-85.
[http://dx.doi.org/10.1016/j.cmet.2013.05.018] [PMID: 23791483]

[14]   Hoang DT, Matsunari H, Nagaya M, *et al.* A conserved rule for pancreatic islet organization. PLoS One 2014; 9(10): e110384.
[http://dx.doi.org/10.1371/journal.pone.0110384] [PMID: 25350558]

[15]   Steiner DJ, Kim A, Miller K, Hara M. Pancreatic islet plasticity: interspecies comparison of islet architecture and composition. Islets 2010; 2(3): 135-45.
[http://dx.doi.org/10.4161/isl.2.3.11815] [PMID: 20657742]

[16]   Caicedo A. Paracrine and autocrine interactions in the human islet: more than meets the eye. Semin Cell Dev Biol 2013; 24(1): 11-21.
[http://dx.doi.org/10.1016/j.semcdb.2012.09.007] [PMID: 23022232]

[17]   Eberhard D, Kragl M, Lammert E. Giving and taking: endothelial and β-cells in the islets of Langerhans. Trends Endocrinol Metab 2010; 21(8): 457-63.
[http://dx.doi.org/10.1016/j.tem.2010.03.003] [PMID: 20359908]

[18]   Goldfine AB, Kulkarni RN. Modulation of β-cell function: a translational journey from the bench to the bedside. Diabetes Obes Metab 2012; 14 (Suppl. 3): 152-60.
[http://dx.doi.org/10.1111/j.1463-1326.2012.01647.x] [PMID: 22928576]

[19]   Bouche C, Lopez X, Fleischman A, *et al.* Insulin enhances glucose-stimulated insulin secretion in healthy humans. Proc Natl Acad Sci USA 2010; 107(10): 4770-5.
[http://dx.doi.org/10.1073/pnas.1000002107] [PMID: 20176932]

[20]   Halperin F, Lopez X, Manning R, Kahn CR, Kulkarni RN, Goldfine AB. Insulin augmentation of glucose-stimulated insulin secretion is impaired in insulin-resistant humans. Diabetes 2012; 61(2): 301-9.
[http://dx.doi.org/10.2337/db11-1067] [PMID: 22275085]

[21]   Rhodes CJ, White MF, Leahy JL, Kahn SE. Direct autocrine action of insulin on β-cells: does it make physiological sense? Diabetes 2013; 62(7): 2157-63.
[http://dx.doi.org/10.2337/db13-0246] [PMID: 23801714]

[22]   Withers DJ, Burks DJ, Towery HH, Altamuro SL, Flint CL, White MF. Irs-2 coordinates Igf-1 receptor-mediated β-cell development and peripheral insulin signalling. Nat Genet 1999; 23(1): 32-40.
[http://dx.doi.org/10.1038/12631] [PMID: 10471495]

[23]   Steiner DF. The proinsulin C-peptidea multirole model. Exp Diabesity Res 2004; 5(1): 7-14.
[http://dx.doi.org/10.1080/15438600490424389] [PMID: 15198367]

[24]   Wahren J, Larsson C. C-peptide: new findings and therapeutic possibilities. Diabetes Res Clin Pract 2015; 107(3): 309-19.
[http://dx.doi.org/10.1016/j.diabres.2015.01.016] [PMID: 25648391]

[25]   Yosten GL, Kolar GR, Redlinger LJ, *et al.* Evidence for an interaction between proinsulin C-peptide and GPR146. J Endocrinol 2013; 218(2): B1-8.
[http://dx.doi.org/10.1530/JOE-13-0203]

[26]   McKillop AM, Ng MT, Abdel-Wahab YH, Flatt PR. Evidence for inhibitory autocrine effects of proinsulin C-peptide on pancreatic β-cell function and insulin secretion. Diabetes Obes Metab 2014; 16(10): 937-46.
[http://dx.doi.org/10.1111/dom.12300] [PMID: 24702738]

[27]   Rorsman P, Berggren PO, Bokvist K, *et al.* Glucose-inhibition of glucagon secretion involves activation of GABAA-receptor chloride channels. Nature 1989; 341(6239): 233-6.
[http://dx.doi.org/10.1038/341233a0] [PMID: 2550826]

[28]   Soltani N, Qiu H, Aleksic M, *et al.* GABA exerts protective and regenerative effects on islet β cell s and reverses diabetes. Proc Natl Acad Sci USA 2011; 108(28): 11692-7.
[http://dx.doi.org/10.1073/pnas.1102715108] [PMID: 21709230]

[29]   Layden BT, Durai V, Newman MV, *et al.* Regulation of pancreatic islet gene expression in mouse islets by pregnancy. J Endocrinol 2010; 207(3): 265-79.
[http://dx.doi.org/10.1677/JOE-10-0298] [PMID: 20847227]

[30]   Rieck S, White P, Schug J, *et al.* The transcriptional response of the islet to pregnancy in mice. Mol Endocrinol 2009; 23(10): 1702-12.
[http://dx.doi.org/10.1210/me.2009-0144] [PMID: 19574445]

[31]   Kim H, Toyofuku Y, Lynn FC, *et al.* Serotonin regulates pancreatic β cell mass during pregnancy. Nat Med 2010; 16(7): 804-8.
[http://dx.doi.org/10.1038/nm.2173] [PMID: 20581837]

[32]   Ohara-Imaizumi M, Kim H, Yoshida M, *et al.* Serotonin regulates glucose-stimulated insulin secretion from pancreatic β cells during pregnancy. Proc Natl Acad Sci USA 2013; 110(48): 19420-5.
[http://dx.doi.org/10.1073/pnas.1310953110] [PMID: 24218571]

[33]   Park KH, Kim BJ, Shawl AI, Han MK, Lee HC, Kim UH. Autocrine/paracrine function of nicotinic acid adenine dinucleotide phosphate (NAADP) for glucose homeostasis in pancreatic β-cells and adipocytes. J Biol Chem 2013; 288(49): 35548-58.
[http://dx.doi.org/10.1074/jbc.M113.489278] [PMID: 24165120]

[34]   Nakashima K, Kanda Y, Hirokawa Y, Kawasaki F, Matsuki M, Kaku K. MIN6 is not a pure β cell line but a mixed cell line with other pancreatic endocrine hormones. Endocr J 2009; 56(1): 45-53.
[http://dx.doi.org/10.1507/endocrj.K08E-172] [PMID: 18845907]

[35]   Ahrén B, Lundquist I. Effects of two cholecystokinin variants, CCK-39 and CCK-8, on basal and stimulated insulin secretion. Acta Diabetol Lat 1981; 18(4): 345-56.
[http://dx.doi.org/10.1007/BF02042819] [PMID: 6277122]

[36]   Lavine JA, Raess PW, Stapleton DS, et al. Cholecystokinin is up-regulated in obese mouse islets and expands β-cell mass by increasing β-cell survival. Endocrinology 2010; 151(8): 3577-88.
[http://dx.doi.org/10.1210/en.2010-0233] [PMID: 20534724]

[37]   Linnemann AK, Neuman JC, Battiola TJ, Wisinski JA, Kimple ME, Davis DB. Glucagon-like peptide-1 regulates cholecystokinin production in β-cells to protect from apoptosis. Mol Endocrinol 2015; 29(7): 978-87.
[http://dx.doi.org/10.1210/me.2015-1030] [PMID: 25984632]

[38]   Layden BT, Angueira AR, Brodsky M, Durai V, Lowe WL Jr. Short chain fatty acids and their receptors: new metabolic targets. Transl Res 2013; 161(3): 131-40.
[http://dx.doi.org/10.1016/j.trsl.2012.10.007] [PMID: 23146568]

[39]   Tang C, Ahmed K, Gille A, et al. Loss of FFA2 and FFA3 increases insulin secretion and improves glucose tolerance in type 2 diabetes. Nat Med 2015; 21(2): 173-7.
[http://dx.doi.org/10.1038/nm.3779] [PMID: 25581519]

[40]   Priyadarshini M, Villa SR, Fuller M, et al. An acetate-specific GPCR, FFAR2, regulates insulin secretion. Mol Endocrinol 2015; 29(7): 1055-66.
[http://dx.doi.org/10.1210/me.2015-1007] [PMID: 26075576]

[41]   McNelis JC, Lee YS, Mayoral R, et al. GPR43 potentiates β cell function in obesity. Diabetes 2015; 64(9): 3203-17.
[http://dx.doi.org/10.2337/db14-1938] [PMID: 26023106]

[42]   Rodriguez-Diaz R, Abdulreda MH, Formoso AL, et al. Innervation patterns of autonomic axons in the human endocrine pancreas. Cell Metab 2011; 14(1): 45-54.
[http://dx.doi.org/10.1016/j.cmet.2011.05.008] [PMID: 21723503]

[43]   Watson E, Hahm S, Mizuno TM, et al. VGF ablation blocks the development of hyperinsulinemia and hyperglycemia in several mouse models of obesity. Endocrinology 2005; 146(12): 5151-63.
[http://dx.doi.org/10.1210/en.2005-0588] [PMID: 16141392]

[44]   Cocco C, Brancia C, Pirisi I, et al. VGF metabolic-related gene: distribution of its derived peptides in mammalian pancreatic islets. J Histochem Cytochem 2007; 55(6): 619-28.
[http://dx.doi.org/10.1369/jhc.6A7040.2007] [PMID: 17312015]

[45]   Petrocchi-Passeri P, Cero C, Cutarelli A, et al. The VGF-derived peptide TLQP-62 modulates insulin secretion and glucose homeostasis. J Mol Endocrinol 2015; 54(3): 227-39.

[http://dx.doi.org/10.1530/JME-14-0313] [PMID: 25917832]

[46]   Stephens SB, Schisler JC, Hohmeier HE, *et al.* A VGF-derived peptide attenuates development of type 2 diabetes *via* enhancement of islet β-cell survival and function. Cell Metab 2012; 16(1): 33-43.
[http://dx.doi.org/10.1016/j.cmet.2012.05.011] [PMID: 22768837]

[47]   Schisler JC, Fueger PT, Babu DA, *et al.* Stimulation of human and rat islet β-cell proliferation with retention of function by the homeodomain transcription factor Nkx6.1. Mol Cell Biol 2008; 28(10): 3465-76.
[http://dx.doi.org/10.1128/MCB.01791-07] [PMID: 18347054]

[48]   Bartolomucci A, Bresciani E, Bulgarelli I, *et al.* Chronic intracerebroventricular injection of TLQP-21 prevents high fat diet induced weight gain in fast weight-gaining mice. Genes Nutr 2009; 4(1): 49-57.
[http://dx.doi.org/10.1007/s12263-009-0110-0] [PMID: 19247701]

[49]   Yang YH, Szabat M, Bragagnini C, *et al.* Paracrine signalling loops in adult human and mouse pancreatic islets: netrins modulate β cell apoptosis signalling *via* dependence receptors. Diabetologia 2011; 54(4): 828-42.
[http://dx.doi.org/10.1007/s00125-010-2012-5] [PMID: 21212933]

[50]   Yang YH, Manning Fox JE, Zhang KL, MacDonald PE, Johnson JD. Intraislet SLIT-ROBO signaling is required for β-cell survival and potentiates insulin secretion. Proc Natl Acad Sci USA 2013; 110(41): 16480-5.
[http://dx.doi.org/10.1073/pnas.1214312110] [PMID: 24065825]

CHAPTER 7

β-Cell Adaptability During Pregnancy

Jens Høiriis Nielsen[1,2,*], **Signe Horn**[1,3], **Jeannette Kirkegaard**[1,3], **Amarnadh Nalla**[1,2] and **Birgitte Søstrup**[1,2]

[1] *Department of Biomedical Sciences, University of Copenhagen, Denmark*

[2] *Centre for Fetal Programming, Copenhagen, Denmark*

[3] *Novo Nordisk A/S, Måløv, Denmark*

Abstract: Pregnancy is a physiological condition associated with β-cell mass expansion occurring in response to increased insulin demand. If the insulin resistance is not compensated by proper augmented insulin production gestational diabetes will occur. As reviewed herein, pregnancy induced hormonal changes have occupied scientists since the beginning of the last century where important discoveries of the hormonal regulation of metabolism during pregnancy have been accomplished. Of the multiple hormonal and metabolic changes the somatolactogenic hormones, placental lactogens (PL) and placental growth hormone (GH-V) are the most described and are found to have dual roles by induction of insulin resistance and promotion of β-cell function and expansion. More recently, the direct effects on isolated pancreatic islets and the influence of signaling pathways involved in the adaptation of β-cell growth and function during pregnancy have been elucidated. This has identified contributions of a number of known peptide hormones and growth factors (EGF, NGF, HGF, IGFs, GLP-1) and steroid hormones (progesterone, estrogens, glucocorticoids). In addition, glucokinase has been found to be essential for the both proliferation and glucose stimulated insulin secretion during pregnancy. Some transcriptional activators and repressors (FoxM1, HNF4α, Myc, Bcl6, Men1) have been implicated in β-cell growth and survival, but also systemic factors like betatrophin, serotonin and osteoprotegerin have been reported to stimulate β-cell proliferation during pregnancy. Gene expression studies and proteomics of islets from pregnant rodent have furthermore revealed upregulation of a number of genes (*e.g.* cyclophilin B, stathmins, dlk-1, trefoil factor-3 and several others) that may influence β-cell growth and function during pregnancy although the mechanisms driving these changes are not yet known. Similarly,

* **Corresponding author Jens Høiriis Nielsen:** Department of Biomedical Sciences, University of Copenhagen, Denmark; Tel: +45 28757721; E-mail: jenshn@sund.ku.dk

David J. Hill (Ed.)

circulating factors in serum from pregnant women have been identified. Among the stimulating factors are peptide fragments of alpha-1 antitrypsin, kininogen-1, apolipoprotein-1, fibrinogen alpha chain and angiotensinogen. An intriguing question remains about the origin of the increased β-cell mass in pregnancy. In humans, studies have primarily reported an increase in the number of small islets, suggesting that neogenesis as the primary driver of β-cell mass expansion in human. In rodents, however, β-cell replication is believed to be the primary mechanism, although increased expression the neogenesis marker, neurogenin-3, has also been reported in pancreas of pregnant rodents. Interestingly, recent studies have suggested that the apparent loss of β-cells occurring during development of diabetes may be due to de-differentiation rather than cell death, suggesting contributions from mechanisms going beyond neogenesis and replication. In summary, gestational diabetes (GDM) is associated with lack of appropriate adaptation of the β-cells that may be due to a reduced pre-pregnancy β-cell mass, lack of stimulating hormones and growth factors or appearance of β cytotoxic metabolites or factors. This chapter reviews the existing knowledge of multiple factors and put forward new mechanisms of pregnancy induced β-cell mass expansion, which are not yet completely understood.

Keywords: Diabetes, Growth hormone, Insulin, Pancreatic β-cells, Pregnancy, Prolactin.

INTRODUCTION

Pregnancy is a unique physiological condition associated with expansion of the β-cell mass in response to an increased insulin demand. As type 2 diabetes (T2D) is characterized by insulin insufficiency due to a lack of compensatory β-cell mass expansion, a better understanding of the mechanisms involved in β-cell adaptation during pregnancy, may help to elucidate the pathogenesis and lead to interventions in the treatment of T2D. Pregnancy is characterized by increased food intake, weight gain, changes in metabolism towards "facilitated anabolism" after food intake and "accelerated starvation" during fasting, in order to maintain optimal supply of nutrients to the fetus [1]. In normal pregnancy, these metabolic changes lead to an increase in the postprandial plasma glucose level which elicits an exaggerated insulin secretion in particular in late pregnancy. The physiological link between pregnancy and the mechanisms failing during development of T2D are further underscored by an increased risk of developing gestational diabetes (GDM). Interestingly, certain genetic forms of diabetes are often diagnosed during pregnancy [2]. As will be described in this chapter the adaptation of the β-cells to pregnancy is a complex process triggered by a series of known, and yet unknown, systemic stimuli. β-cell adaptation is accomplished by increased β-cell function such as enhanced glucose stimulated insulin secretion (GSIS), as well as, by the formation of new β-cells most likely by a combination of replication and neogenesis. The changes in β-cell mass during normal and diabetic conditions are illustrated in Fig. (**1**) [3].

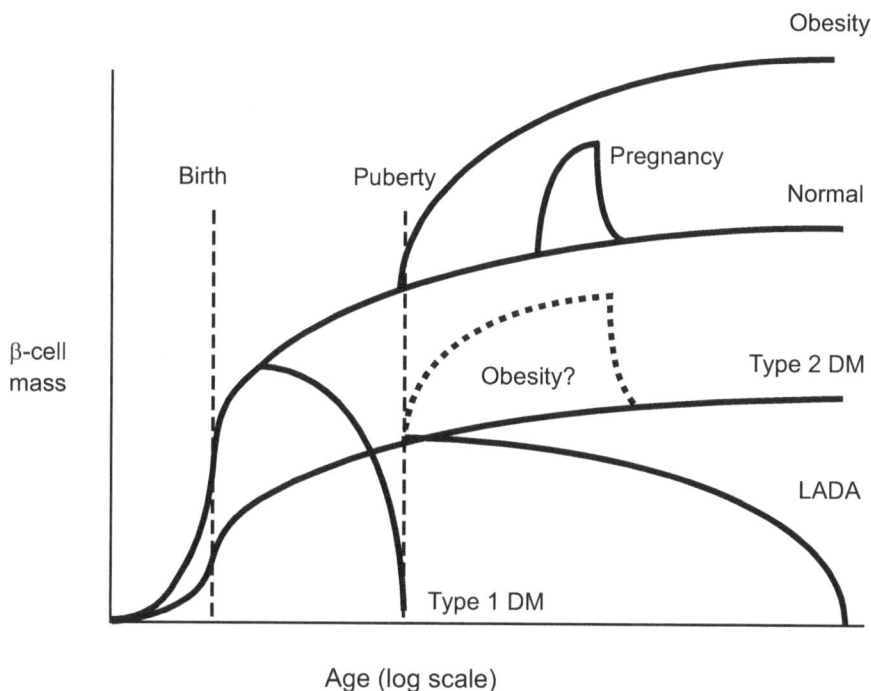

Fig. (1). Schematic illustration of changes in β-cell mass in response to increased insulin demand *i.e.* perinatal period, intrauterine growth retardation, obesity, pregnancy and type 1 diabetes (DM), type 2 diabetes (DM) and Latent Autoimmune Diabetes in Adults (LADA or type 1½ diabetes) [3].

Morphological Changes

Changes in the morphology of the endocrine pancreas during pregnancy in both humans and animals were described already in the early 20th century [4]. In 1930, Akehi [5] found hypertrophy, proliferation and newly formed islets during pregnancy in rabbits, which was followed by rapid atrophy and reduction of β-cells after parturition. In the same year, Macleod [6] described a diabetic woman that had a marked reduction in her insulin requirement in the last trimester, suggesting a temporary recovery of β-cells. In the following decades, several animal studies reported β-cell mass expansion in response to pregnancy. In 1933 Cramer [7] described the occurrence of numerous very small islets in pregnant rats, suggesting that new islets are formed. Hellman [8], however, described an increase in the ratio between β and α cells in pregnant rats, suggesting replication of existing β-cells. Later on, studies have suggested, that the β-cell mass expansion occurs by a combination of both neogenesis and replication. Hellerström found a 25% increase in total islet volume in pregnant mice [9] and Van Assche *et al.* reported a doubling of the endocrine mass with an increase in both small and large islets in pregnant rats [10 - 12]. In addition, these authors found an increase in the number of glucagon, somatostatin and PP cells [10 - 12].

Although more scarce, studies of β-cell adaption during pregnancy in humans, also report adaptive β-cell mass expansion. In 1932 Rosenloecher reported an increase in islet number and β-cell hyperplasia based on a study of 12 pregnant women [13]. In 1978, in a study on five pregnant women, van Assche *et al.* likewise found a 2-fold increase in the endocrine pancreas with an increase in β-cell content of around 10 per cent [14]. More recently, Butler *et al.* analyzed tissue sections from 18 pregnant women and found 1.4-fold increase in β-cell area, as well as, an increase in islet number [15]. However, the authors did not observe an increase in islet size or replicating β-cells suggesting that β-cell neogenesis could be the major contributor of islets expansion in human pregnancy.

Hormonal Changes During Pregnancy

As schematized in Fig. (**2**) [16], pregnancy is associated with changes in the secretion of several hormones that have both direct and indirect impact on the regulation of the pancreatic hormones. Some hormones are exclusively produced in the placenta and are naturally upregulated during pregnancy such as chorionic gonadotropin (CG), placental lactogen(s) (PL), placental growth hormone (GH-V). Others are produced by the mother and can be both up- and downregulated. Growth hormone (GH) produced by the pituitary gland is downregulated, whereas pituitary prolactin (PRL) is upregulated. Pregnancy is also associated with an increase in the gonadal hormones such as estrogens and progesterone, as well as, adrenal hormones such as glucocorticoids. The secretion of adipokines is changed during pregnancy. Although leptin is increased it does not suppress appetite because of hypothalamic leptin resistance. Adiponectin is reduced and TNFα is increased, which may contribute to the insulin resistance [16, 17]. Despite the facts that the hormonal changes are generally observed across multiple species, some species differences have been reported. Rat and mouse *e.g.* have two distinct PLs whereas ovine PL has been reported to show more GH-like properties [18]. Furthermore, several PRL-like molecules such as *e.g.* the proliferins have been reported in the placenta and other tissues [19].

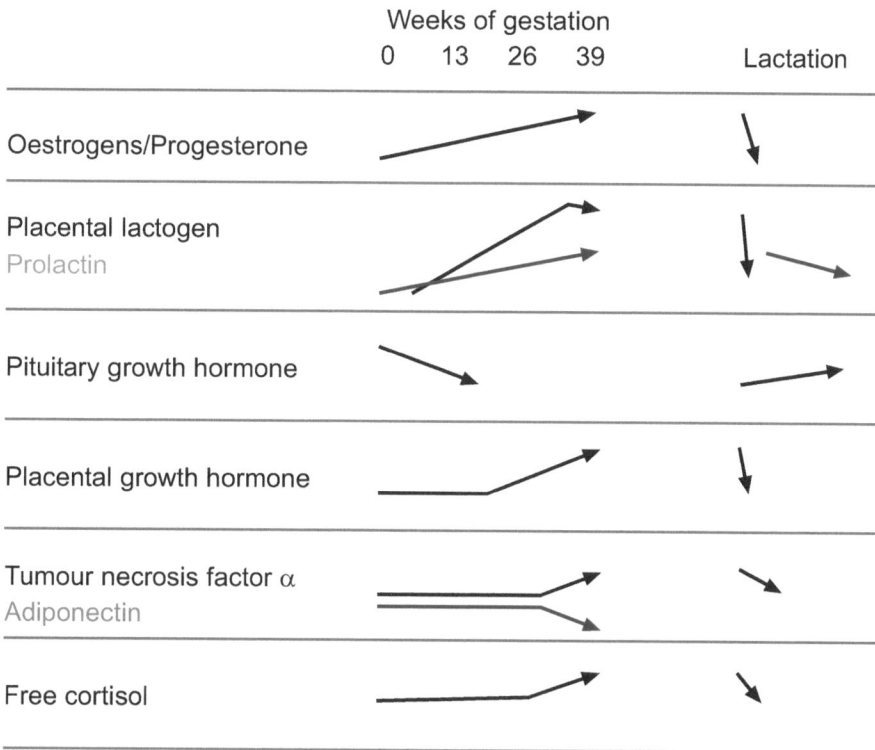

Fig. (2). Schematic illustration of hormonal changes during pregnancy and lactation in humans. During pregnancy estrogens, progesterone, placental lactogen and prolactin are increasing. Pituitary growth hormone is decreasing whereas placental growth hormone is increasing. Tumor necrosis factor α is increased whereas adiponectin is decreased in late pregnancy. Free cortisol is increased in late pregnancy. The placental hormones decline after parturition whereas prolactin remains elevated and the other hormones return to the non-pregnancy levels [16].

Somatolactogenic Hormones – Systemic Effects

The earliest comment on an association between the pituitary and diabetes was by Loeb who in 1884 mentioned that glucosuria often occurs in cases of pituitary tumors [20]. The role of pituitary hormones in glucose homeostasis was further studied the 1920'es and 30'es when Houssay and collaborators demonstrated increased insulin sensitivity in hypophysectomized dogs [21]. The authors, furthermore, reported diabetogenic effects when pituitary extracts were administered to dogs [21]. These effects mimicked the observation made on dogs treated with purified GH [22]. Interestingly, there is a marked species and age difference in the diabetogenic effects observed after administration of pituitary extracts or purified GH. Temporal diabetes was obtained in adult dogs, cats, ferrets, monkeys, rabbits and goats [23]. However, GH was only found to be diabetogenic in rats after partial pancreatectomy [24]. Strikingly, pituitary extracts were neither diabetogenic in puppies and kittens nor in pregnant and lactating animals [23], suggesting that GH is not diabetogenic under physiological conditions with active growth and sufficient compensatory supply of insulin to overcome the insulin resistance.

Also in humans some discrepancy about the effects of GH exists. Although GH has been found to be diabetogenic, it is able to stimulate GSIS in hypophysectomized patients [25]. Development of diabetes in acromegaly may only occurs in patients with decreased GSIS [26]. Although the diabetogenic effects of GH in patients with acromegaly are often reversible after cessation of the treatment, or after tumor resection in acromegaly, it has been reported that prolonged GH treatment of partial pancreatectomized dogs can produce permanent diabetes with "ballooning degeneration" of β-cells [27]. Insulin or phlorizin treatment can reverse diabetogenic effects in GH-induced diabetic cats [28], however, insulin treatment is not crucial for survival, as GH-induced diabetic dogs can be kept alive for a long time even without insulin treatment [29]. Together these results suggest that GH-induced diabetes is secondary to the hyperglycemia that causes degranulation, exhaustion and eventually death of the β-cells. Further complicating the interpretation, pituitary extracts and GH have also been reported to have anti-diabetogenic effects. Thus, in the early stages of treatment, pituitary extracts have been reported to increase mitotic activity in the islets of dogs [30]. Similar effects were observed after systemic administration in rats [29] and rabbits [31].

Using another approach Martin *et al.* observed marked β-cell hyperplasia when studying the effects of a transplantable GH- and PRL-producing tumor (MtT-W15) on insulin secretion in rats [32]. Treatment with prolactin induced an increase in the insulin content in rats [30]. However, prolactin has been shown to induce transitory or permanent diabetes in dogs and cats [21]. Likewise, PRL treatment induced mild glucose intolerance but not diabetes in dogs [33]. In human, hyperprolactinemia leads to development of mild glucose intolerance and hyperinsulinemia [34].

A possible explanation for the contradictory effect observed by the administration of pituitary hormones could be that the hormones have opposite dose-dependent effects. In support of this, Park *et al.* found that low dose PRL administration increased β-cell regeneration and insulin secretion, whereas, high dose administration stimulated β-cell regeneration but impaired insulin secretion in 90% pancreatectomized rats probably due to an exaggerated insulin resistance [35].

Somatolactogenic Hormones – Lessons from *In Vitro* Studies

Development of techniques to study islet function *in vitro,* such as whole pancreas or islet perfusion, as well as, islet and tissue culture has enabled studies on islets responses during direct exposure to hormones and nutrients.

Exploiting these techniques several studies on pregnancy-induced changes in islets have been reported. Malaisse *et al.* reported improved GSIS from pancreatic tissues isolated from rats at day 15 and 20 of pregnancy. These effects could, furthermore, be mimicked in hypophysectomized rats treated with hPL [36]. Similar positive effects were reported by Costrini and Kalkhoff in 1971 [37], as well as, by Bone and Taylor who reported improved insulin biosynthesis due to an increase in cAMP levels [38]. In addition, Green and Taylor [39] observed increased glucose sensitivity, as well as, increased islet size at day 19 of pregnancy in rats. Interesting, dietary carbohydrate restriction prevented this effect, suggesting that the increased glucose intake is involved [40].

Suggesting a direct role and further supporting a positive role of pituitary hormones on islets, islets isolated from hypophysectomized rats were found to contain and release less insulin compared to islets isolated from non-hypophysectomized, an effect which could be partly rescued by administration of GH [41]. Later studies showed, that treating islets isolated from hypophysectomized rats with GH lead to a small increase in insulin biosynthesis [42]. Whittaker and Taylor, as well as, Pierluissi *et al.* likewise found an increase in both insulin release and biosynthesis after treating rat islets for either 16 h or 4 day with rat GH [43 - 45].

Human GH, ovine PRL and human PL have also been shown to have direct effects on DNA synthesis in islets isolated from neonatal rats and to sustain the insulin release to the culture medium in mouse islets cultured for 14d in RPMI1640 supplemented with 0.5% newborn calf serum (Fig. **3**) [46]. In contrast, bGH showed no effect on DNA synthesis when tested on monolayer cultures of neonatal rat pancreas [47]. The lack of effects could, however, be due to the use of 10% fetal calf serum (FCS) that already may contain PRL or GH or other growth factors. In support of this, increased mitotic activity was found in the same culture system using only 0-0.1% FCS [48]. Long term cultures of rat islets or monolayers of dissociated islet cells with hGH resulted in a 10-20-fold increase in the number of β-cells [49]. This increase depends on the activation of both the GH receptor (GHR) and the PRL receptor (PRLR) which bind hGH in rat insulinoma cells [50]. Sorenson and coworkers studied the effect of PRL on islets isolated from rats and found a decrease in the threshold for GSIS, as well as, an increased coupling among the β-cells [51, 52]. They found that glucokinase, hexokinase and Glut-2 were upregulated during pregnancy and that these changes could be reproduced *in vitro* by PRL treatment [53, 54]. In addition, the authors observed an increase in cell proliferation, which correlated with the secretion of the placental lactogens in pregnant rats (Fig. **4**) [55]. By comparing the mitogenic effects of GH, PRL and PL they found that PRL and PL were more potent than GH both in rat, mouse and human islets [56]. In rat islets, PL-II has been shown to be a more potent stimulator of insulin secretion compared to PRL and GH [57].

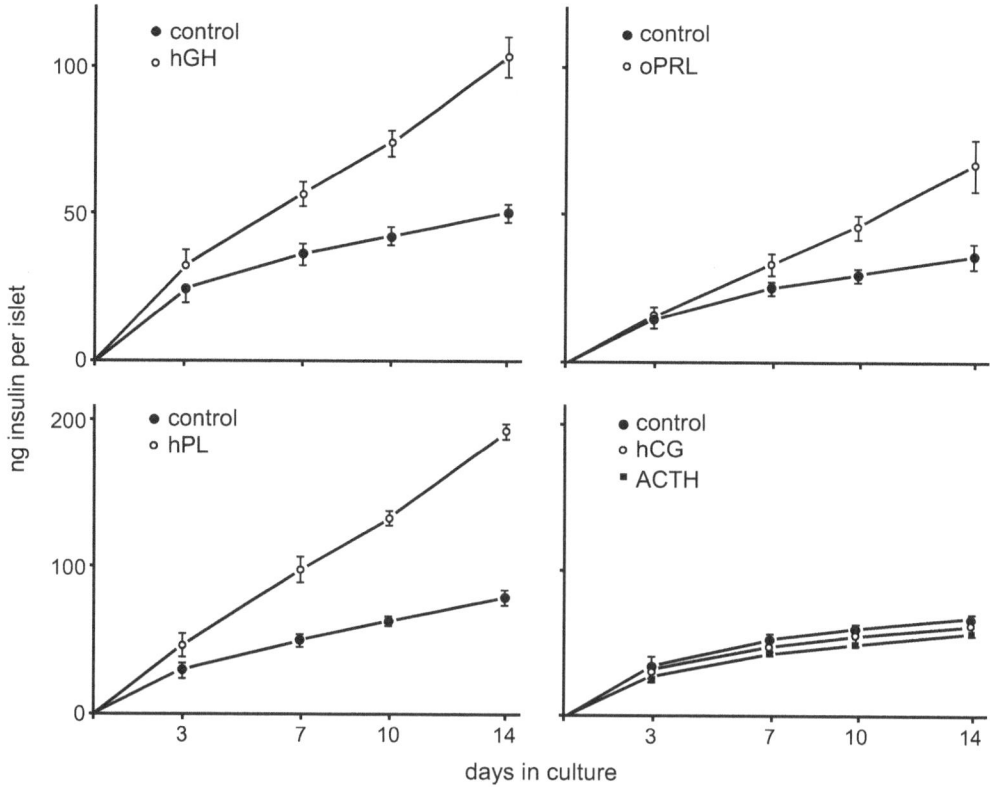

Fig. (3). Effect of hGH, oPRL, hPL, hCG and ACTH (1 µg/ml) in insulin release from adult mouse islets culture medium RPMI 1640 containing 0.5% newborn calf serum [46].

There is still some disagreement with regard to the effect of the somatolactogenic hormones on human β-cell proliferation. Brelje and Sorenson found that islets from a female donor at age 16 responded to hPRL, hGH and hPL while islets from a female donor at age 49 had a low response to hPRL and hPL [56]. A male donor at age 17 responded to hPRL and hPL whereas islet from several older donors had no response to the hormones [58]. In our own studies, we found very low levels of GHR and PRLR mRNA in islets from most donors except for two males at age 6 and 8 and one female at age 40 (who may have been pregnant but this could not be confirmed) (Nielsen, unpublished). Although one study has reported a 2-fold increase in β-cell proliferation in human islet cells cultured with hPRL [59], most studies find no or very little β-cell proliferation in human islets *in vitro* [60, 61].

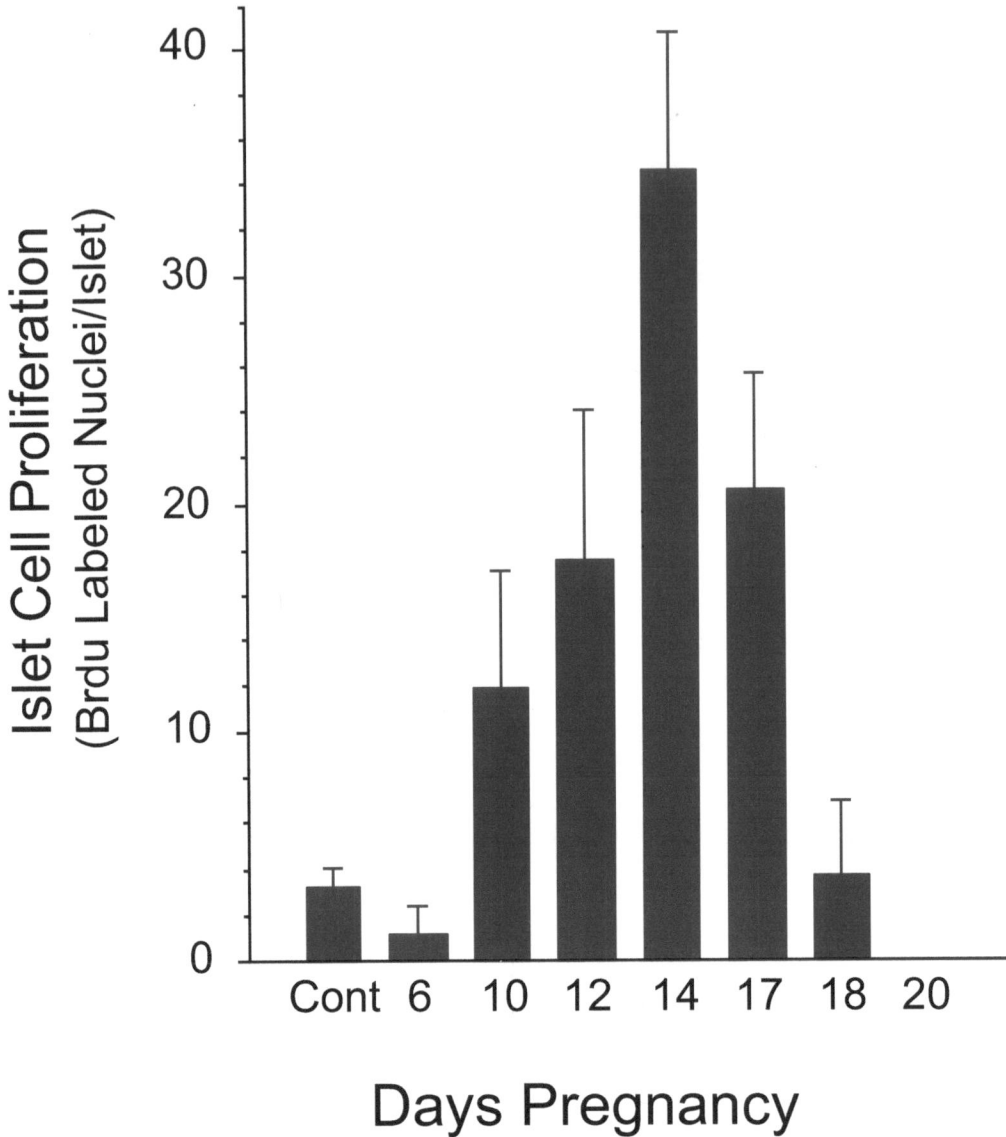

Fig. (4). Proliferation of islet cells during pregnancy in rats [55].

Somatolactogenic Hormones - Lessons from Transgenic Mice

The role of lactogenic hormones in β-cell development, maturation and function has been emphasized in studies using transgenic mice. Transgenic mice with β-cell specific overexpression of hPL, which binds to the PRLR, resulted in increased β-cell proliferation and hypoglycemia [62]. Opposite, mice carrying a deletion of the PRLR had reduced β-cell mass and GSIS [63] suggesting that PRLR agonists are involved in embryonic development of the β-cells. In fact, we have detected GHR and PRLR mRNA expression in the early embryonic mouse pancreas although their functions remain to be studied [64]. As PRLR null mice cannot carry a pregnancy beyond midgestation heterozygous females have been investigated [65]. They showed a reduction of β-cell mass and proliferation, as well as, impaired insulin secretion and glucose tolerance. Transgenic mice overexpressing hGH showed an increase in both islet number and islet size suggesting both increased proliferation and neogenesis [66]. Opposite, GHR knockout results in dwarfism and a marked decrease in β-cell mass and islet size. As expected there was an increased peripheral sensitivity to insulin resulting in reduced blood glucose levels [67]. In order to study the role of GHR specifically in β-cells, a β-cell specific knock-out mouse was generated by LeRoith and collaborators [68]. These mice had normal β-cell mass when fed chow diet but showed impaired β-cell growth and glucose tolerance in response to high-fat diet. Unfortunately, the effect of pregnancy was not reported in this study. On the other side, it has been argued that the PL and GH-V are not required for β-cell adaptation to pregnancy based on the occurrence of pregnancy in women whit deletion mutations in these genes, however, this could be due to compensation by pituitary PRL and GH [17] or some of the prolactin-like molecules [19]. In order to identify the role of placental signals in the islet adaptation to pregnancy the effect of pseudo-pregnancy on β-cell proliferation has recently been studied [69]. Although there was no change in glucose homeostasis, pseudo-pregnancy was found to increase β-cell proliferation, suggesting a role of non-placental factors in the increased β-cell mass which may include pituitary PRL or PRL-like molecules.

Somatolactogenic Hormones – Mechanism of Action

PRL has been shown to bind to rat islets [71, 72] and expression of both GHR and PRLR has been detected in rat insulinoma cells [50]. A study on GH and PRL receptor expression in the rat pancreas revealed a pregnancy induced increase in both PRLR and GHR mRNA (Fig. **5**) [70]. Similarly, PRLR and GHR were increased in the liver but with the interesting differences that pancreas from males expressed PRLR mRNA whereas the liver did not and that the pancreas only expressed the long form of the PRLR whereas the liver also expressed a short form. Sorenson and collaborators [73] demonstrated the presence of PRL receptors in both β and α cells and confirmed their upregulation during pregnancy. Interestingly, Møldrup and coworkers [70] found that the expression of the PRLR was stimulated by both PRL and GH, whereas the GHR was stimulated by glucocorticoids but not by GH and PRL. Therefore the effects of the lactogenic hormones become increasingly important during pregnancy as PL, PRL, GH and GH-V stimulate PRLR expression. These data, furthermore, suggest that, the perpetuating stimulation of β-cell proliferation observed in long term cultures stimulated with hGH, mentioned above, is most likely mediated by a continuous increase in PRLR expression. This is supported by the studies of Brelje and Sorenson [74] who cultured neonatal rat islets for 36 days in the presence of rPRL or rGH and found a much higher increase in both proliferation and insulin production with rPRL than with rGH.

As GH is known to induce IGF-1 in the liver and other tissues it has been investigated if the effect of GH on the islets is mediated by IGF-1. However, IGF-1 mRNA was not detected in neonatal rat islets exposed to hGH. Furthermore, addition of IGF-1 or antiserum to IGF-1 did not have any influence of the GH-induced proliferation [75]. In contradiction, there are other reports on expression of IGF-1 in the fetal human and rat endocrine pancreas [76 - 78] and both α and β-cells express IGF-1 receptors [79]. Similarly, β-cell specific knock out of IGF-1R results in impaired β-cell function and glucose intolerance [80]. However, no change in β-cell mass was observed, which could suggest a role of IGF-1 in β function and not in growth. More recently, it was reported that GH and IGF-1 act synergistically on proliferation of β-cells in isolated mouse islets [81]. The authors proposed a model where GHR and IGF-1R physically interact with each other and activate both STAT and AKT signaling. The discrepancy with our results in rat islets [75] may due to the species as mouse islets are more dependent on the presence of serum factors *e.g.* IGF-1 than rat islets in order to maintain the β-cell function during *in vitro* culture [82].

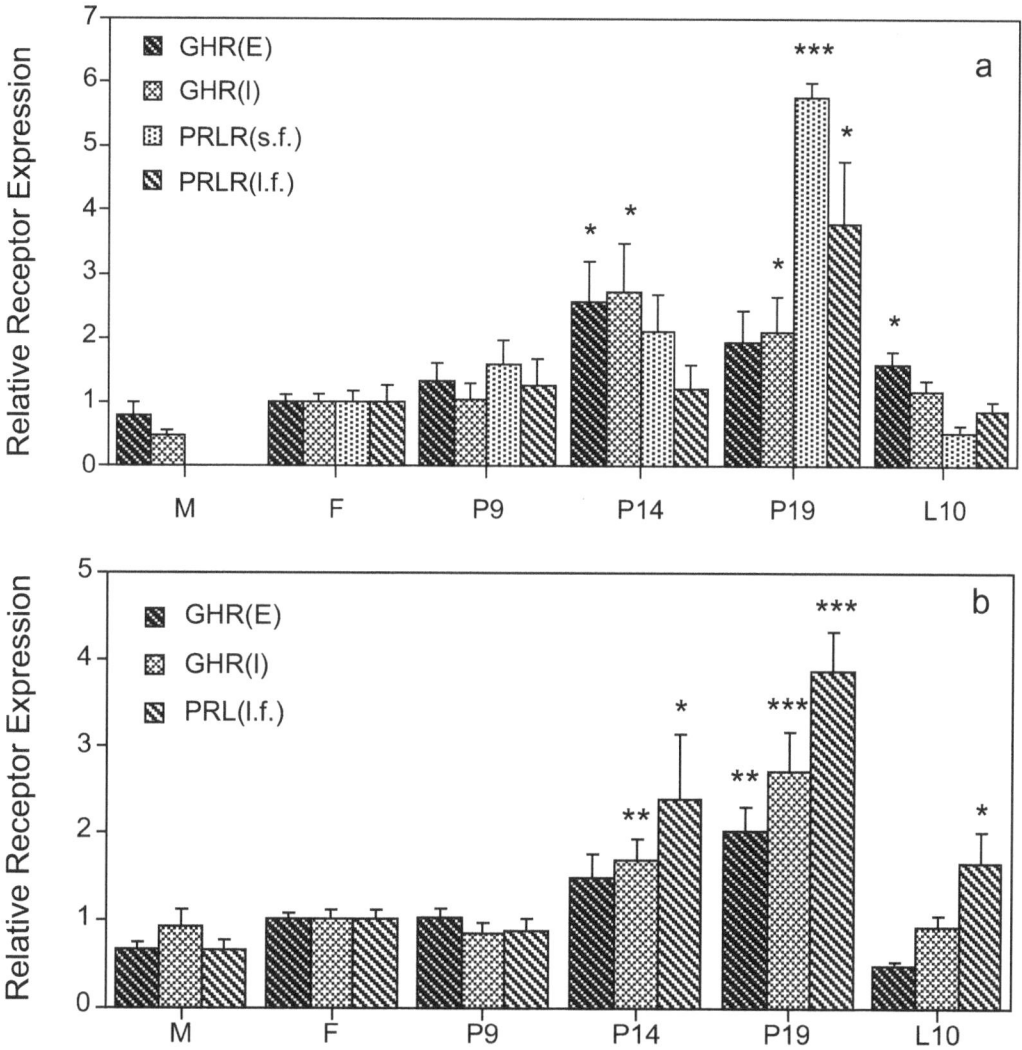

Fig. (5). Effect of gender, pregnancy and lactation on GH and PRL receptor mRNA expression in rat liver (panel a) and pancreas (panel b).Total RNA was extracted from liver (5 µg) and pancreas (50 µg) of male (M), female (F), pregnant female (P) and lactating (L) rats were subjected to analysis of GHR and PRLR mRNA by RNase protection assay. The values are given relative to the values in normal female rats. Both short form corresponding to the extracellular domain (E) and the long form corresponding to the intracellular domain (I) of GHR and short form (s.f.) and long form (l.f.) of PRLR were measured. * $p<0.05$, **$p<0.01$, ***$p<0.001$ [70].

Somatolactogenic Hormones – A Downstream Role of STAT5

The PRLR and GHR belong to the cytokine receptor family that activates the

JAK/STAT/SOCS (Janus kinase/signal transducers and activators of transcription/suppressors of cytokine signaling) pathway. PRLR and GHR activate the JAK2 tyrosine kinase that phosphorylates the transcription factor STAT5 which regulates several target genes in β-cells [83]. The mitogenic effect of hGH in rat insulinoma cells (INS-1) can be inhibited by a dominant negative STAT5 mutant (DN-STAT5) [84] and overexpression of a constitutive active STAT5 mutant (CA-STAT5) in INS-1 cells or neonatal rat islets can stimulate proliferation in the absence of hGH by increasing the expression of cyclin D2 that harbors a STAT5 binding site in its promoter [85]. Brelje *et al.* [86] demonstrated a difference in the kinetics of activation of STAT5B by rGH and rPRL with PRL showing a longer duration than GH. The response to PRL was inhibited by a short pre-treatment with PRL or GH, whereas, the response to GH was inhibited with any length of pre-treatment with GH but not with PRL. There are several possible explanations for these differences. One is the selective upregulation of the PRLR by PRL, PL and GH *via* the STAT5 binding site in the promoter [87] and another is the possible difference in induction of SOCS proteins that act as negative feed-back inhibitors of both GH and PRL signaling [88]. Overexpression of SOCS3 in INS-1 cells reduced hGH-induced proliferation by inhibiting DNA binding of both STAT5 and STAT3 [89]. *In vivo* overexpression of SOCS3 in β-cells resulted in reduced β-cell mass but an enhanced glucose tolerance in the female mice [90]. Interestingly SOCS3 overexpression in INS-1 cells also protected against the apoptotic effects of the proinflammatory cytokines IL-1β and IFNγ [91] and thus may act as survival factor. In addition hGH protected INS-1 cells against the cytotoxic cytokines *via* STAT5 activation of the anti-apoptotic BclX$_L$ [92]. Similar results were reported for the protective effect of PRL on streptozotocin and dexamethasone induced β-cell death [93]. Jackerott *et al.* [94] investigated transgenic mice expressing either DN- or CA-STAT5 in β-cells after treatment with multiple low doses of streptozotocin, as a model of type 1 diabetes, and high fat diet feeding, as a model of type 2 diabetes. The severity of both forms of diabetes was enhanced in the mice with reduced STAT5 (DN-STAT5) activity, whereas the mice with increased STAT5 (CA-STAT5) activity had less severe diabetic symptoms. In conclusion, these results suggest that high STAT5 activity in β-cells may mediate both β-cell adaptation during pregnancy and protect β-cells against functional impairment by inflammatory cytokines like TNFα that is increased during pregnancy. However, the effect of pregnancy was not analyzed in these animals. In addition to the JAK2/STAT5 pathway activation GHR and PRLR can also activate ERK1/2, adenylate cyclase/cAMP and intracellular calcium-accumulation [95]. The stimulation of extracellular calcium uptake is independent of the JAK/STAT pathway, as a GH receptor mutant unable to bind JAK2 but with an intact C-terminal domain was shown to mediate the GH-induced calcium uptake [96]. In a microarray study of PRL treated rat islet Boschero *et al.* [97] found upregulation of glucokinase, STAT3, ERK1, cyclin D2, CDK4, IGF2R, somatostatin and TRH confirming the involvement of additional pathways and factors (see below).

Role of Glucocorticoid Hormones

As mentioned above, pregnancy is associated with marked increases in steroid hormone levels, *i.e.* estrogens, progesterone and glucocorticoids (Fig. **2**). The influence of adrenal hormones has been known since 1909, where Porges observed hypoglycemia in patients with Addison's disease which could be reproduced by adrenalectomy in dogs [98]. In 1960 is was furthermore found that administration of glucocorticoid hormones can lead to hyperglycemia and diabetes [21]. The diabetogenic effects are, however, associated with β-cell hyperplasia and restoration of the GSIS in adrenalectomized rats [30, 99, 100]. In our laboratory we found a biphasic effect of hydrocortisone on insulin secretion in mouse islets cultured for 3 weeks with a maximal stimulation at 100 nM [101]. However, in agreement with other studies [47] we found that hydrocortisone inhibited DNA synthesis in neonatal rat islets [102]. The concentration of glucocorticoid hormone increases markedly during pregnancy. As it has been shown that dexamethasone can inhibit the PRL stimulated β-cell proliferation *in vitro*, it is possible that the glucocorticoids may be involved in the decline in β-cell proliferation in late pregnancy which occurs in spite of the high concentration of the somatolactogenic hormones [103]. However, a positive effect of glucocorticoids on β-cells in pregnancy may be implied by the stimulation of GH receptor expression in islets *in vitro* [70].

Role of Progesterone

In perfused rat pancreas Sutter-Dub [104] found that the biphasic GSIS was markedly increased at pregnancy day 20 and that ovariectomized rats treated with estradiol showed enhanced GSIS whereas progesterone treatment had no effect, suggesting that only estradiol has a direct stimulatory effect on the pancreas. However, isolated islets from rats treated with estradiol and progesterone or progesterone alone have been reported to show increased GSIS and response to theophylline that is not dependent of activation of adenylate cyclase [105]. The effect of progesterone on theophylline stimulated GSIS could be reproduced *in vitro* after culture for 20 hours [105]. Similar results were obtained after culture of mouse islets for 14 days where there was an increase in GSIS but a decrease in the insulin content of the islets [106]. Progesterone (1 µg/ml) was found to inhibit DNA synthesis in neonatal rat islets *in vitro* but the combination of hPL and progesterone (0.1 µg/ml) did not prevent the hPL induced increase in insulin and DNA contents in mouse islets [106]. An inhibition of PRL-induced β-cell proliferation by progesterone (1 µg/ml) was also reported by Sorenson and colleagues in 1993 [107]. Interestingly, 14 days culture of mouse islets in 10% serum from women pregnant in third trimester resulted in 50% increased insulin release, 50% decrease in insulin content and 20% increase in DNA content compared to islets cultured in 10% serum from non-pregnant women, suggesting contributions from factors increased in the circulation during pregnancy [106]. Treatment of rats with progesterone *in vivo* was shown to stimulate proliferation of both β and alpha cells in pregnant rats [108, 109]. However, it is not yet clear if progesterone is involved in the declined β-cell proliferation in late pregnancy.

Role of Estrogens

Already in the in the 1940ies, a role of estrogens in diabetes was identified [21]. Thus, 95% pancreatectomy resulted in diabetes in 89% of the males only 27% of the females became diabetic [110]. Removal of the testes decreased, whereas removal of the ovary increased, the incidence. Initially, estrogens aggravated diabetes but several months' treatment with estrogens or corticoids was protective whereas testosterone increased the incidence as did high doses of progesterone. Estradiol was found to induce hypertrophy and hyperplasia in the islets and β-cells as well as neogenesis [110, 111]. Ovariectomized rats have increased food intake and higher fasting glucose levels and lower insulin levels than normal rats when pair-fed which could be reversed by estradiol treatment. This was reflected in the improved GSIS in islets isolated from estradiol treated rats. These islets also showed increase binding of estradiol and progesterone suggesting an important role of these hormones in the β-cells [112]. In order to test if the sex steroids affected the islet cell replication, isolated rat islets were cultured for 22h and the incorporation of tritiated thymidine was measured. Increased DNA synthesis was observed in islets from ovariectomized rats, normal rats fed with glucose, pregnant rats at day 12 and 19 and islets pretreated with glucose for 24h *in vitro*, whereas inhibition was seen by addition of progesterone, serum from pregnant and normal rats, amino acid supplement and IBMX. No effect was seen by addition of estradiol or rGH to 5% fetal calf serum [113]. It was concluded that the steroid hormones were not responsible for the increased β-cell replication during pregnancy but there was a striking correlation between the blood glucose levels and the DNA synthesis *in vitro*. Estradiol has been shown to enhance GSIS both *in vivo* and *in vitro via* a membrane bound receptor [114, 115] and to protect β-cells from apoptosis and to prevent streptozotocin diabetes in mice [116]. Emerging evidence for a role of non-coding RNAs in β-cell adaptation to pregnancy is being reported. In islets from pregnant rats there was a decrease in miR-338-3p expression which could be mimicked by activating the G protein-coupled estrogen receptor GPR30 and the GLP-1R that act *via* the cAMP/PKA pathway. Blocking miR-338-3p resulted in increased β-cell proliferation and survival *in vitro* and *in vivo* [117]. Thus, estrogens may contribute to the adaptation of the β-cells to pregnancy *via* both nuclear and membrane bound receptors.

Role of Epidermal Growth Factor (EGF)

The role of EGF was investigated using a dominant negative EGF receptor under the control of the Pdx-1 promotor. Otonkoski and collaborators [118] showed that these mice lost their ability to expand the β-cell mass during pregnancy and found that the PL had no effect on β-cell proliferation on the isolated islets from these mice [119]. EGFR signalling mainly exerts its biological effects *via* the PI3K/AKT/mTOR pathway and Ras/Raf/MAPK pathway [120]. In support of this, mTOR signalling promotes the cell proliferation in adult mice [121] and pregnant mice dosed with mTOR inhibitor rapamycin blunted the adaptive β-cell proliferation [122].

Role of Nerve Growth Factor (NGF)

By transcriptional analysis of islets from pregnant mice Kaestner *et al.* [123] found a dramatic increase in the expression of the NGF receptor suggesting that NGF contribute to β-cell proliferation and/or survival. NGF has previously been shown play an important role in the embryonic development of the endocrine pancreas [124].

Role of Hepatocyte Growth Factor (HGF)

The role of HGF is underscored by the fact that pregnant mice have increased circulating hepatocyte growth factor (HGF) levels [125] and increased islet expression of HGF receptor (c-Met). Mice lacking pancreatic HGF receptors fail to expand their maternal β-cell mass [125]. In various rodent models of insulin resistance, a consistent correlation between circulating HGF levels and the compensatory mechanism was found. Furthermore systemic administration of HGF increased β-cell mass and replication in a dose-dependent manner in mice [126]. Interestingly, EGFR induced PI3K/AKT/mTOR signalling increases HGF receptor expression in murine mammary cells isolated at midgestation [127], and it is thus tempting to speculate that EGFR signalling also stimulate β-cell HGF receptor expression in an autocrine/paracrine manner during pregnancy. Further linking HGF to replication, a prominent increase in islet endothelium HGF expression was reported to correlate with the peak in β-cell replication in pregnant rats [128].

Role of Survivin

Several labs, including ours, have found an upregulation of survivin RNA (also known as Birc5) in islets of pregnant mice. Survivin is viewed as a crucial β-cell survival factors since it inhibits caspases [129, 130] and since young mice lacking survivin progressively loss their β-cells [131]. A thorough evaluation of the pregnant RIP-Cre:survivin [fl/fl] mice revealed that survivin is not only a β-cell survival factor, but also drives proliferation as the mutant mice displayed impaired adaptive β-cell replication, as well as glucose intolerance [132]. Supportive of a both proliferative and survival function of survivin its expression closely mirrored the expression profile of Ki67 during compensatory β-cell mass expansion [123]. How survivin expression is ensured in β-cells during pregnancy is not completely clear. Two studies point to EGFR signaling as an upstream inducer of survivin: First, decreased EGFR activity was shown to attenuate pregnancy-induced survivin expression [119] and secondly, EGF stimulation of islets enhanced survivin protein stability *in vitro* [133]. PRL receptor signaling also stimulated survivin expression in islets [119] and its stimulation was blocked by inhibitors of the PI3K/AKT/mTOR, the JAK2/STAT5, and the Ras/Raf/MAPK pathways [119, 132], thus several upstream pathways contribute to survivin induction.

Role of IGF-Binding Protein 5 (Igfbp5)

We and several other labs [123, 134, 135] have reported increased IGF-binding protein 5 (Igfbp5) mRNA expression in islets as a response to pregnancy. We have confirmed a significant increase in Igfbp5 protein synthesis rate resulting in a 5-fold increased Igfbp5 protein expression in islets at midgestation. Interestingly, β-cell specific expression of an activated form of AKT revealed Igfbp5 as an AKT target gene in β-cells [132, 136]. Thus, it is plausible that pregnancy-stimulated AKT signalling upregulates Igfbp5 expression in β-cells as a response to pregnancy. Igfbp5 was found to induce cell replication in non-β-cell tissue [137] and it is upregulatated in several tumors including pancreatic islet adenocarcinomas [138]. Igfbp5 binds and modulate IGF 1 and IGF 2 activity [139], but it also possess effects independent of IGFs [140]. Mice lacking Igfbp5 became glucose intolerant upon diet-induced obesity pointing towards a role in β-cell mass expansion [136]. Igfbp5$^{-/-}$ mice can, however, still increase their β-cell mass in response to pregnancy [136] which might be a result of compensation by other Igfbp family members, such as Igfbp3, which was found significantly increased in serum of adult mice lacking Igfbp5 [141].

Role of Glucagon-like Peptide 1 (Glp-1)

Glp-1 is a well-known growth factor for rodent β-cells and activates cyclin D1 *via* multiple pathways including cAMP/PKA, PI3K and MAP kinases [142] and has been shown also to depend on EGF receptor activation [143]. Recently Flatt *et al.* [144] found an increased production of Glp-1 and GIP in the alpha cells in pregnant mice and demonstrated that mice lacking the receptor for Glp-1 abolished the adaptation of the β-cells to pregnancy and even reduced the insulin content suggesting that Glp-1 but not GIP contribute to the β-cell expansion during pregnancy due to intra-islet Glp-1. The stimulating effect of Glp-1 on GSIS depends primarily on the activation of the cAMP/PKA pathway. As mentioned above this pathway is augmented in pregnancy, which in part may be due to GH/PRL induced upregulation of the $G\alpha_s$ protein (Carlsson, unpublished). In addition, GH/PRL upregulates the A-kinase anchoring protein 18 (AKAP18 or AKAP 7/15) that has been shown to increase the GSIS response to Glp-1 [145].

Role of Glucose and Fatty Acids

Pregnancy is associated with increased food intake and it has been suggested that nutrients are the most important stimuli for the increased β-cell mass in pregnancy. As mentioned above studies by Green and Taylor in 1972 [39] showed that pair feeding of pregnant rats and non-pregnant rats reversed the functional changes and Nieuwenhuizen *et al.* showed that food restriction or insulin treatment of pregnant rats prevented the increased β-cell proliferation in spite of the elevated PL secretion [146, 147]. As there is a decline in the glucose levels close to term [148] it may well be that the decline in β-cell proliferation in late pregnancy reflects this condition. Recent studies support the important role of glucose metabolism initiated by glucokinase in β-cell proliferation [149]. As mentioned above lactogenic hormones activates glucokinase in β-cells [53] and synthetic glucokinase activators have been shown to stimulate β-cell replication in mice *via* activation of IRS-2 and Pdx-1 [150]. Interestingly, patients with neonatal hypoglycemia due to activating mutations in the glucokinase gene have islet hyperplasia and increased β-cell proliferation as well as apoptosis [151]. It is therefore pertinent to unravel the mechanisms that determine balance between glucose-induced β-cell proliferation and β-cell toxicity. Increased fatty acid levels have been suggested to be involved in the decreased β-cell proliferation in late pregnancy but surprisingly palmitate and oleate have been shown to stimulate β-cell proliferation in particular in combination with PRL [152] and may thus contribute to the increased β-cell mass during pregnancy.

Role on Neurogenin 3 (Ngn-3)

As mentioned above several studies of pancreas during pregnancy have found evidence for neogenesis of islets by differentiation of putative progenitor cells in the exocrine pancreas [153]. Ngn-3 plays a crucial role in the embryonic development of the endocrine pancreas (for a review see) [154]. We have detected Ngn-3 immunoreactivity in the pancreas of pregnant mice [155, 156] localized to the pancreatic ducts, acini and islets suggesting that neogenesis occurs in rodents, although proliferation of existing β-cells is the major contributor [155]. As mentioned above, this could be opposite in human, where β-cell proliferation seems to be very limited whereas numerous small islets have been observed indicating neogenesis [15]. In our study, we found that serum from pregnant women but not hGH could induce expression of Ngn-3 in fetal rat pancreas *in vitro* [156]. Contradictory, other studies showed no increase in the expression of Ngn3 mRNA in the pancreas from pregnant mice [157] and no evidence for neogenesis of β-cells in pregnant mice [158]. Thus, the role of Ngn-3 in β-cell neogenesis during pregnancy needs further investigation.

Role of Forkhead Box Protein M1 (FoxM1)

The transcription factor Forkhead box protein M1 (FoxM1) is also directly increased by prolactin [159] and FoxM1 is in general upregulated in proliferating cells including neonatal pancreatic endocrine cells [160]. Pregnant mice with a pancreatic deletion of FoxM1 displayed a striking reduction of β-cell proliferation leading to profound glucose intolerance, however, a reduction in β-cell mass of female knockout mice, already prior to pregnancy, complicates the interpretation of the gestational effects [159]. The functions of FoxM1 on β-cell mass may derive partly by its direct regulation of survivin [161].

Role of Hepatic Nuclear Factor 4α (Hnf4α)

During pregnancy the transcription factor Hnf4α may indirectly contribute to EGFR stimulated Ras/Raf/MAPK activation in part by down-regulation of the tumor suppressor gene suppression of tumorigenicity 5 (ST5). Pregnant mice lacking Hnf4α in β-cells displayed lower EGF receptor mediated ERK phosphorylation and a blockage in compensatory β-cell mass growth [162]. Activated ERK phosphorylates the transcription factor Myc proto-oncogene protein [163], a potent driver of β-cell replication and growth in adult β-cells. We found a 3-fold upregulated of Myc at the mRNA level in islets at midgestation [164]. Importantly, ectopic β-cell expression of Myc forced cell cycle entry, but the end result is massive β-cell loss and diabetes [165]. This is consistent with the notion that appropriate β-cell mass expansion rely on mechanisms ensuring survival of replicating β-cells as discussed by [166]. This might also be what is visualized in pancreatic section from Type 2 diabetes patients, where apoptotic β-cells are often found together two and two suggesting that they fail to survive just after mitosis [166, 167].

Role of B-cell Lymphoma 6 Protein Homolog (Bcl6) and Menin (Men1)

Another phospho-STAT5 target gene is B-cell lymphoma 6 protein homolog (Bcl6) which was found to be upregulated in islets from pregnant mice [164]. It is a transcriptional repressor which both anti-apoptotic and proliferative effects in β-cells [93, 168]. The proliferative effect of Bcl6 was first described by Karnik *et al.* who provided evidence for Bcl6 repression of Menin 1 (Men1). Through its association with the trithorax complex, Men1 acts as a transcriptional activator of cell cycle inhibitors such as p18 and p27 [168]. Thus, prolactin receptor induced Bcl6 expression blocks Men1 expression, which subsequently reduces the levels of the cell cycle blockers p18 and p27 [3].

Role of Betatrophin

Recently Melton *et al.* [169] identified angiopoietin-like protein 8 (ANGPTL8) as a potential growth factor for β-cells based on its upregulation in fat and liver in mouse that were rendered insulin resistant by treatment with an insulin receptor antagonist. The factor was called betatrophin as it was able to stimulate β-cell proliferation, β-cell mass and improve glucose tolerance. It is upregulated 20-fold in the liver during pregnancy and it may be speculated to contribute to the increased β-cell mass during pregnancy. However, more studies are needed before its role can be determined.

Role of Osteoprotegerin (OPG)

Recently, a proliferative role of the soluble decoy receptor osteoprotegerin (OPG) was revealed by Kondegowda *et al.* [170]. OPG is increased in islets of pregnant mice [123] and in the circulation of pregnant women [171]. It is regulated by prolactin, most likely *via* STAT5 [172], and is an unusual member of the Tumor Necrosis Factor Receptor Superfamily since it lacks a transmembrane domain, and thus is secreted. Secreted OPG bind and occupy the protein receptor activator of NF-kB ligand (RANKL) which therefore is prevented from stimulating the RANKL/RANK pathway. Both RANKL and RANK are expressed in human islets [173] and Kondegowda *et al.* showed that the RANKL/RANK pathway is a brake for β-cell replication in both human and mice [170]. Thus, OPG is one of only a few proteins that has been shown to increase human β-cell replication *via* its inhibition of the RANKL/RANK pathway. Evaluation of β-cells in diabetes patients treated for osteoporosis with the drug, Denosumab, could clarify the therapeutic potential of OPG since this osteoporosis drug likewise bind and inhibits the RANKL/RANK pathway.

Role of Serotonin

Several differential gene expression studies in islets from pregnant mice have identified the enzymes involved in serotonin synthesis tryptophan 5-hydroxylase 1 (Tph1) and tryptophan 5-hydroxylase 2 (Tph2) as highly upregulated during pregnancy [123, 135, 164, 174]. Lactogenic signalling increases islet expression of these two rate-limiting enzymes responsible for the conversion of tryptophan to serotonin. PRLR signalling *via* a combination of JAK2/STAT5, Ras/Raf/MAPK and PI3K/AKT/mTOR [175] is at least partly responsible for the strong increase of Tph1 and Tph2 expression, which results in a dramatic increase in islet serotonin production [135, 174]. Kim *et al.* demonstrated that dietary restriction of tryptophan in pregnant mice arrested islet serotonin production, and as a consequence the pregnant mice became severely glucose-intolerant. Additionally, serotonin deficient pregnant mice completely lacked adaptive β-cell growth [135] and the positive effect on β-cell proliferation of serotonin is evolutionary conserved to at least zebrafish [176, 177]. Kim *et al.* further suggested that the link between serotonin and β-cell proliferation was to be found in an increased islet expression of the serotonin receptor Hrt2b, whose activity drives expression of several G1 cyclin genes in hepatocytes [178]. Although the suggestion of such autocrine/paracrine induced proliferation is biological appealing, we and others [119, 123, 134, 174] have not been able to confirm upregulation of Hrt2b on either the RNA level or protein level in islets of pregnant mice [164]. Recently, serotonin was also found to increase glucose-stimulated insulin secretion by lowering of the β-cell glucose threshold during pregnancy *via* stimulation of the serotonin-gated ion channel Hrt3. This caused a mild sodium influx decreasing the resting membrane potential, and thus increased the β-cell membrane excitability [179]. Despite appropriate pregnancy induced β-cell mass expansion, pregnant Hrt3$^{-/-}$ mice had lower GSIS compared to wild type pregnant mice and they were in general less glucose tolerant [179].

Role of Cyclophilin B

In our proteomic screen of islets from pregnant mice we identified an upregulation of peptidyl-prolyl cis-trans isomerase B (Ppib, also known as Cyclophilin B) [164] that has been associated with cellular survival, proliferation and JAK2/STAT5-independent prolactin signalling. Notably, Ppib was found to co-immunoprecipitate with prolactin [180]. Based on localization studies and sequence analysis, Rycyzyn *et al.* suggested that Ppib-associated prolactin, binding to the prolactin receptor, caused the receptor to undergo endocytosis in a complex with Ppib-bound prolactin. Subsequently, the nuclear translocation signal of Ppib was found to facilitate nuclear entrance of prolactin, and enhanced nuclear prolactin-induced proliferation up to 9-fold *in vitro* [180]. However, whether such events occur *in vivo* is not known. Additionally, Ppib is induced upon ER-stress, where it prevents cell death [180, 181].

Role of Stathmins

We have identified stathmin and stathmin2 protein as significantly upregulated in islets from pregnant mice [164]. The two proteins are known to play a growth-promoting role in pancreatic cancer [182]. Additionally, stathmin2 is a selective marker for regenerating axons following nerve injury [183], and the similarity between neurons and β-cells suggests a potential role for the stathmins, not only in β-cell mass expansion, but possibly also in β-cell regeneration.

Role of Delta-like 1 (dlk-1)

β-cells have been shown to produce dlk-1 (also known as preadipocyte factor-1 or Pref-1) [184]. The protein contains 6 EGF-like motifs and can be converted to a soluble form FA-1 (fetal antigen-1) that is abundant in amniotic fluid and in the circulation during pregnancy [106]. It is highly expressed in islets around birth and during pregnancy and is upregulated by hGH in isolated rat islets. The function is unknown but it is speculated that it may induce dedifferentiation that permits proliferation of β-cells [185].

Role of Trefoil Factors (TFF)

Trefoil factors (TFFs) 1, 2 and 3 are small peptides that are predominantly found in the gastrointestinal tract and other mucosal tissues where they are involved in the maintenance and protection of surface integrity. They are also expressed in the β-cells in the pancreas [186]. TFF3 is most abundant in rat and human islets and in some duct cells. hGH stimulates the expression of TFF3 in isolated rat islets and TFF3 promoted the attachment of the cells but not the proliferation and GSIS [185] whereas other studies have suggested that TFF3 is involved in β-cell proliferation [187]. Interestingly the circulating levels in serum are dramatically increased in pregnant women [188] but the function is still not known.

Other Differentially Expressed Islet Genes

In addition to the factors mentioned above genome wide screening for differentially expressed genes at the transcriptional and posttranslational levels have revealed up- and down regulation of numerous genes. Among the most upregulated proteins in pregnant mouse islets, we [164] identified a large group of proteins including secretogranin-1 (Chgb), matrillin-2 (Matn2), peroxisomal bifunctional enzyme (Ehhahd), isovaleryl-CoA dehydrogenase (Ivd), bone morphogenetic protein 1 (Bmp1), Itga6 protein (Itga6), Ezrin (Ezr) and 4-trimethylaminobutyraldehyde dehydrogenase (Aldh9A1), which have all previously been demonstrated upregulated at the mRNA level in islets in response to pregnancy [123, 134, 135, 174]. The evaluation of the functional roles of these proteins in the adaptation of the β-cells to pregnancy requires further investigations.

Circulating Factors in Pregnancy that may Influence β-Cell Function

Another approach to identify factors that influence β-cell growth and function during pregnancy is to apply proteomics and metabolomics to blood or amniotic fluid and use functional screening on isolated islets or β-cell lines. We have fractionated peptides from serum from pregnant women and tested the fractions for DNA synthesis in rat insulinoma cells [189]. In the stimulatory fractions we identified peptide fragments from alpha-1 antitrypsin (AAT), kininogen-1, apolipoprotein A-1, fibrinogen alpha chain and angiotensinogen. AAT is elevated in pregnancy and has previously been shown to stimulate insulin secretion and protect β-cells against proinflammatory cytokines [190]. In order to assess the role of PL in the stimulatory effect of pregnancy serum we used a PRLR receptor antagonist [191] and found in preliminary experiments that it blocked the mitogenic effect in neonatal rat islets [155], confirming that PRLR agonists are the major mitogenic factor(s) in the circulation although it does not exclude an additional effect of other factors.

Diabetes in Pregnancy

Pregnancy is a risk factor for development of diabetes. Up to 15% of pregnant women develop gestational diabetes [192] and it is conceivable that it is due to lack of compensatory increase in β-cell mass and function to overcome the peripheral insulin resistance [193]. Obesity is a risk factor for development of GDM and women with pre-existing T1D or T2D will most often require intensive insulin therapy in order to maintain normoglycemia. As mentioned above there have been reported cases where pregnancy has lead to an improvement of the diabetic condition by the increased β-cell function due to the pregnancy hormones that may compensate for the insulin demand. Thus, in some pregnant women with T1D an improved C-peptide level has been reported [194] which may be due to increased β-cell regeneration or suppression of the autoimmunity [195]. PRL treatment has been shown to prevent autoimmune diabetes in NOD mice [196] and in multiple low dose streptozotocin diabetic mice [197] by enhancing the Th2 lymphocyte response [198]. Women with GDM have a 7-fold increased risk of developing T2D [199]. The mechanisms involved in the development of GDM are not clarified and it does not seem to be associated with changes in the pregnancy hormones [200]. Most probably it is an imbalance between insulin resistance and insulin delivery. Thus a reduction in the β-cell mass by partial pancreatectomy or low dose streptozotocin treatment suggest that the pre-pregnancy β-cell mass is a determining factor [193]. The β-cell mass depends on genetic and environmental factors during early life that influence the epigenetic programming of the endocrine pancreas [201]. Alternatively there may be circulating factors like abnormal metabolites, toxins from food or microbiota that impair the function or viability of the β-cells. For example Wheeler *et al.* [202] have recently identified a furan fatty acid metabolite (CMPF) in serum from women with gestational diabetes that impaired β-cell function *in vitro* and induced glucose intolerance in mice. Instead of loss of β-cells by metabolic stress Accili *et al.* have presented evidence for an alternative mechanism *i.e.* dedifferentiation. The transcription factor FoxO1 is essential for maintaining a normal β-cell number and function and ablation of FoxO1 in β-cells result in hyperglycemia and reduced β-cell mass in response to physiological stress such as multiparity and ageing. However, lineage tracing have shown that loss of β-cells may not be due to cell death but to dedifferentiation with reappearance of progenitor/stem cell markers as Ngn-3, Oct4, Nanog and L-myc [203]. In their model the healthy β-cells have FoxO1 located in the cytoplasm but under metabolic stress it was translocated to the nucleus to maintain the β-cell fate. Under chronic stress FoxO1 declines and β-cell specific genes are turned off and replaced by progenitor specific genes or other islet hormone genes. If this model holds then challenge will be to reactivate the β-cell differentiation program.

β-Cell Adaptation Postpartum

After delivery there is a rapid reduction of the β-cell mass both in lactating and non-lactating rats [148] associated with the disappearance of the β-cell stimulating placental hormones and reduction in the expression of PRLR and GHR (Fig. **5**) [70]. Bonner-Weir and collaborators have demonstrated that the involution of the islets is associated with apoptosis of the β-cells and observed an increased expression of TGFβ1 in rats 3 days postpartum [204]. It would be interesting to investigate if dedifferentiation is involved in this process. Lower blood glucose and insulin levels and lower β-cell mass were found in lactating compared to non-lactating rats which may be due to the higher energy demand of the offspring [148]. In contrast, lactating mice were found to have a higher rate of β-cell proliferation than non-lactating mice [69] which may be due to the higher levels of PRL. The discrepancy may be due to differences in the time period of lactation. Higher intensity of lactation of women after GDM was associated with improved fasting glucose and lower insulin levels at 6-9 weeks postpartum indicating increased insulin sensitivity that may reduce the risk of developing T2D [205].

CONCLUSION

The β-cell plasticity during pregnancy has many similarities to the processes failing during development of type 2 diabetes. There is a marked increase in the requirement for insulin associated with increased food intake and peripheral insulin resistance mainly due to somatolactogenic hormones produced by the placenta. If the demand for insulin is not compensated by an increase in the insulin production hyperglycemia and diabetes will occur. The increased insulin production is accomplished by an increase in the β-cell mass either by proliferation of the existing β-cells or by recruitment of new β-cells from putative progenitor cells in the pancreas. In some species including rodents proliferation is prevailing whereas in other species including humans there is some evidence for neogenesis – although this is still controversial. As discussed in this review numerous hormones and factors are involved in the expansion of the β-cell proliferation but little is known about the mechanisms involved in neogenesis. Some of the signalling pathways are depicted in Fig. (**6**). The lack of human tissue makes it urgent to perform studies in species other than rodents. The classical studies in dogs indicate that during certain periods in life characterized by growth requirement *i.e.* childhood, pregnancy and lactation the organism can compensate for the increased energy demand. Thus, there may exist permissive factors that sensitize β-cells or progenitors to growth and differentiation factors. Future research may show whether some of the already identified factors mentioned in this chapter have such properties.

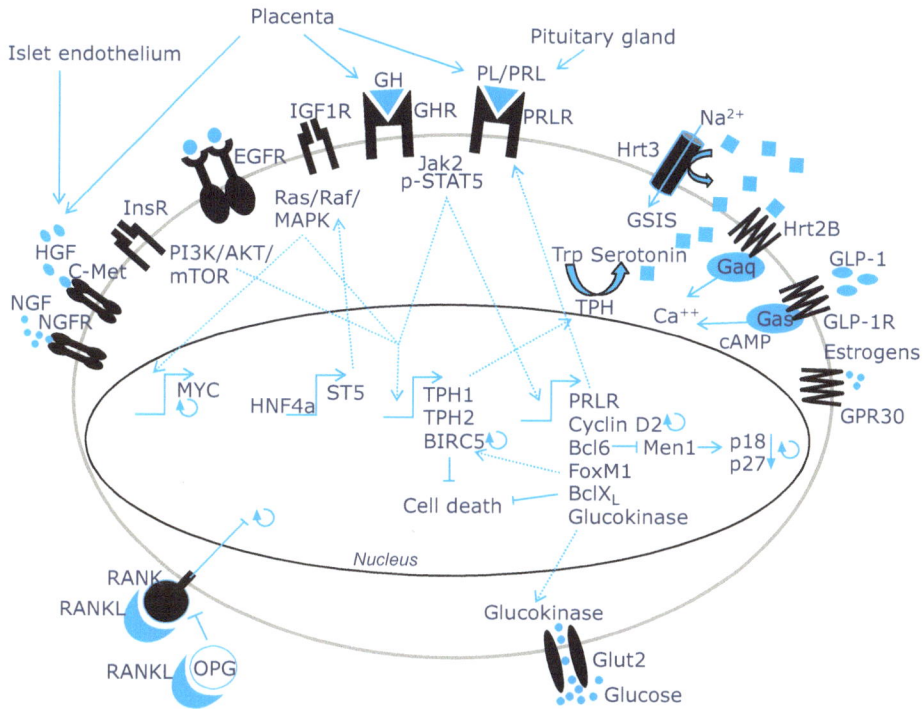

Fig. (6). Schematic illustration of extracellular factors, receptors, signaling pathways and transcription factors involved in β-cell adaptation to pregnancy. References are given in the text.

CONFLICT OF INTEREST

The authors confirm that they have no conflict of interest to declare for this publication.

ACKNOWLEDGEMENTS

The authors want to acknowledge the following funding agencies for support to the studies performed in their laboratories: Danish Diabetes Academy supported by Novo Nordisk Foundation, Novo Nordisk Foundation, Juvenile Diabetes Research Foundation, European Union 6th Frame Work Program, European Foundation for the Study of Diabetes, Danish Research Council for Health Sciences, Novo Nordisk A/S, Danish Research Council for Strategic Research supporting Centre for Fetal Programming and Danish Stem Research Center, and Danish Ministry for Higher Education and Science.

REFERENCES

[1]   Freinkel N. Banting Lecture 1980. Of pregnancy and progeny. Diabetes 1980; 29(12): 1023-35.
        [http://dx.doi.org/10.2337/diab.29.12.1023] [PMID: 7002669]

[2]   Chakera AJ, Spyer G, Vincent N, Ellard S, Hattersley AT, Dunne FP. The 0.1% of the population with
        glucokinase monogenic diabetes can be recognized by clinical characteristics in pregnancy: the
        Atlantic Diabetes in Pregnancy cohort. Diabetes Care 2014; 37(5): 1230-6.
        [http://dx.doi.org/10.2337/dc13-2248] [PMID: 24550216]

[3]   Lee YC, Nielsen JH. Regulation of β cell replication. Mol Cell Endocrinol 2009; 297(1-2): 18-27.
        [http://dx.doi.org/10.1016/j.mce.2008.08.033] [PMID: 18824066]

[4]   Sirtori C. Sul contego delle isole del Langerhans in gravidanza ed puerperio. Ann Ostet Ginecol 1907;
        29: 433-50.

[5]   Akehi T. Internal secretion of the pancreas and the female genital function Part I. Histological
        investigation of the pancreas in pregnancy and puerperium, especially the Langerhans's islands. J Obst
        Gynec 1930; 13: 427-32.

[6]   Macleod JJ. Diabetes as a physiological problem. Br Med J 1930; 1: 868.
        [http://dx.doi.org/10.1016/S0140-6736(01)09257-1]

[7]   Cramer W. Changes in the islets of Langerhans in pregnancy and in other conditions. Q J Exp Physiol
        1933; 23: 127-30.
        [http://dx.doi.org/10.1113/expphysiol.1933.sp000590]

[8]   Hellman B. The islets of Langerhans in the rat during pregnancy and lactation, with special reference
        to the changes in the B/A cell ratio. Acta Obstet Gynecol Scand 1960; 39: 331-42.
        [http://dx.doi.org/10.3109/00016346009159930] [PMID: 14400898]

[9]   Hellerstroem C. The influence of pregnancy and lactation on the endocrine pancreas of mice. Acta Soc
        Med Ups 1963; 68: 17-28.
        [PMID: 14104761]

[10]   Aerts L, Vercruysse L, Van Assche FA. The endocrine pancreas in virgin and pregnant offspring of
        diabetic pregnant rats. Diabetes Res Clin Pract 1997; 38(1): 9-19.
        [http://dx.doi.org/10.1016/S0168-8227(97)00080-6] [PMID: 9347241]

[11]   Van Assche FA. Quantitative morphologic and histoenzymatic study of the endocrine pancreas in
        nonpregnant and pregnant rats. Am J Obstet Gynecol 1974; 118(1): 39-41.
        [PMID: 4271795]

[12]   Van Assche FA, Gepts W, Aerts L. Immunocytochemical study of the endocrine pancreas in the rat
        during normal pregnancy and during experimental diabetic pregnancy. Diabetologia 1980; 18(6): 487-
        91.
        [http://dx.doi.org/10.1007/BF00261705] [PMID: 6106615]

[13]   Rosenloecher K. Die Veränderungen des Pankreas in der Schwangerschaft bei Nensch und Tier. Arch
        Gynakol 1932; 151: 567-75.
        [http://dx.doi.org/10.1007/BF01701915]

[14]   Van Assche FA, Aerts L, De Prins F. A morphological study of the endocrine pancreas in human

pregnancy. Br J Obstet Gynaecol 1978; 85(11): 818-20.
[http://dx.doi.org/10.1111/j.1471-0528.1978.tb15835.x] [PMID: 363135]

[15] Butler AE, Cao-Minh L, Galasso R, *et al*. Adaptive changes in pancreatic β cell fractional area and β cell turnover in human pregnancy. Diabetologia 2010; 53(10): 2167-76.
[http://dx.doi.org/10.1007/s00125-010-1809-6] [PMID: 20523966]

[16] Freemark M. Regulation of maternal metabolism by pituitary and placental hormones: roles in fetal development and metabolic programming. Horm Res 2006; 65 (Suppl. 3): 41-9.
[http://dx.doi.org/10.1159/000091505] [PMID: 16612113]

[17] Newbern D, Freemark M. Placental hormones and the control of maternal metabolism and fetal growth. Curr Opin Endocrinol Diabetes Obes 2011; 18(6): 409-16.
[http://dx.doi.org/10.1097/MED.0b013e32834c800d] [PMID: 21986512]

[18] Forsyth IA. Variation among species in the endocrine control of mammary growth and function: the roles of prolactin, growth hormone, and placental lactogen. J Dairy Sci 1986; 69(3): 886-903.
[http://dx.doi.org/10.3168/jds.S0022-0302(86)80479-9] [PMID: 3519707]

[19] Soares MJ, Konno T, Alam SM. The prolactin family: effectors of pregnancy-dependent adaptations. Trends Endocrinol Metab 2007; 18(3): 114-21.
[http://dx.doi.org/10.1016/j.tem.2007.02.005] [PMID: 17324580]

[20] Loeb M. Ein erklärungsversuch der verschiedenartigen temperaurverhältnisse bei der tuberculösen basilameningitis. Dtsch Arch Klin Med 1883-84; 34: 433-50.

[21] Houssay BA. Other hormones. In: Williams RH, Ed. Diabetes. New York: Paul B. Hoeber 1960; pp. 233-56.

[22] Evans HE, Meyer K, Simpson ME. Disturbance of carbohydrate metabolism in normal dogs injected with the hypophyseal growth hormone. Proc Soc Exp Biol Med 1932; 29: 857-61.
[http://dx.doi.org/10.3181/00379727-29-6114]

[23] Young FG, Korner A. Growth hormone. In: Williams RH, Ed. Diabetes. New York: Pail B. Hoeber 1960; pp. 216-32.

[24] Young FG. Growth hormone and metabolism. New York: Recent Progress in Hormone Research 1953; pp. 471-510.

[25] Luft R, Cerasi E. Effect of human growth hormone on insulin production in panhypopituitarism. Lancet 1964; 2(7351): 124-6.
[http://dx.doi.org/10.1016/S0140-6736(64)90130-8] [PMID: 14160547]

[26] Luft R, Cerasi E, Hamberger CA. Studies on the pathogenesis of diabetes in acromegaly. Acta Endocrinol 1967; 56(4): 593-607.
[PMID: 5630505]

[27] Lazarus SS, Volk BW. Pancreatic adaptation to diabetogenic hormones. AMA Arch Pathol 1959; 67(4): 456-67.
[PMID: 13636627]

[28] Lukens FD, Dohan FC. Morphological and functional recovery of the pancreatic islands in diabetic cats treated with insulin. Science 1940; 92(2384): 222-3.
[http://dx.doi.org/10.1126/science.92.2384.222] [PMID: 17743855]

[29] Richardson KC, Young FG. The pancreotropic action of anterior pituitary extracts. J Physiol 1937; 91(3): 352-64.
[http://dx.doi.org/10.1113/jphysiol.1937.sp003564] [PMID: 16994939]

[30] Lazarus SS, Volk BW. The Pancreas in Human and Experimental Diabetes. New York: Grune & Stratton 1962.

[31] Ogilvie RF. The endocrine pancreas in human and experimental diabetes. In: Cameron MP, O'Connor M, Eds. Aetiology of diabetes mellitus and its complications Ciba Foundation Colloquia on Endocrinology. London: J & A Churtill 1964; pp. 49-66.
[http://dx.doi.org/10.1002/9780470719350.ch4]

[32] Martin JM, Akerblom HK, Garay G. Insulin secretion in rats with elevated levels of circulating growth hormone due to MtT-W15 tumor. Diabetes 1968; 17(11): 661-7.
[http://dx.doi.org/10.2337/diab.17.11.661] [PMID: 4879849]

[33] Houssay BA, Anderson E, Bates RW, Li CH. Diabetogenic action in prolactin. Endocrinology 1955; 57(1): 55-63.
[http://dx.doi.org/10.1210/endo-57-1-55] [PMID: 13251205]

[34] Landgraf R, Landraf-Leurs MM, Weissmann A, Hörl R, von Werder K, Scriba PC. Prolactin: a diabetogenic hormone. Diabetologia 1977; 13(2): 99-104.
[http://dx.doi.org/10.1007/BF00745135] [PMID: 852641]

[35] Park S, Kim DS, Daily JW, Kim SH. Serum prolactin concentrations determine whether they improve or impair β-cell function and insulin sensitivity in diabetic rats. Diabetes Metab Res Rev 2011; 27(6): 564-74.
[http://dx.doi.org/10.1002/dmrr.1215] [PMID: 21557442]

[36] Malaisse WJ, Malaisse-Lagae F, Picard C, Flament-Durand J. Effects of pregnancy and chorionic growth hormone upon insulin secretion. Endocrinology 1969; 84(1): 41-4.
[http://dx.doi.org/10.1210/endo-84-1-41] [PMID: 4882000]

[37] Costrini NV, Kalkhoff RK. Relative effects of pregnancy, estradiol, and progesterone on plasma insulin and pancreatic islet insulin secretion. J Clin Invest 1971; 50(5): 992-9.
[http://dx.doi.org/10.1172/JCI106593] [PMID: 4928265]

[38] Bone AJ, Taylor KW. Mitabolic adaptation to pregnancy shown by increased biosynthesis of insulin in islets of Langerhans isolated from pregnant rat. Nature 1976; 262(5568): 501-2.
[http://dx.doi.org/10.1038/262501a0] [PMID: 785279]

[39] Green IC, Taylor KW. Effects of pregnancy in the rat on the size and insulin secretory response of the islets of Langerhans. J Endocrinol 1972; 54(2): 317-25.
[http://dx.doi.org/10.1677/joe.0.0540317] [PMID: 4560943]

[40] Green IC, Taylor KW. Insulin secretory response of isolated islets of Langerhans in pregnant rats: effects of dietary restriction. J Endocrinol 1974; 62(1): 137-43.
[http://dx.doi.org/10.1677/joe.0.0620137] [PMID: 4604621]

[41] Martin JM, Gagliardino JJ. Effect of growth hormone on the isolated pancreatic islets of rat *in vitro*. Nature 1967; 213(5076): 630-1.
[http://dx.doi.org/10.1038/213630a0] [PMID: 5340262]

[42]   Sun AM, Lin BJ, Haist RE. Studies on the effects of growth hormone and thyroxine on proinsulin synthesis and insulin formation in the isolated islets of Langerhans of the rat. Can J Physiol Pharmacol 1972; 50(12): 1147-51.
[http://dx.doi.org/10.1139/y72-167] [PMID: 4569899]

[43]   Whittaker PG, Taylor KW. Direct effects of rat growth hormone in rat islets of langerhans in tissues culture. Diabetologia 1980; 18(4): 323-8.
[http://dx.doi.org/10.1007/BF00251014] [PMID: 6998803]

[44]   Pierluissi J, Pierluissi R, Ashcroft SJ. Effects of growth hormone on insulin release in the rat. Diabetologia 1980; 19(4): 391-6.
[http://dx.doi.org/10.1007/BF00280526] [PMID: 7000602]

[45]   Pierluissi J, Pierluissi R, Ashcroft SJ. Effects of hypophysectomy and growth hormone on cultured islets of Langerhans of the rat. Diabetologia 1982; 22(2): 134-7.
[http://dx.doi.org/10.1007/BF00254843] [PMID: 6277717]

[46]   Nielsen JH. Effects of growth hormone, prolactin, and placental lactogen on insulin content and release, and deoxyribonucleic acid synthesis in cultured pancreatic islets. Endocrinology 1982; 110(2): 600-6.
[http://dx.doi.org/10.1210/endo-110-2-600] [PMID: 6276141]

[47]   Chick WL. β cell replication in rat pancreatic monolayer cultures. Effects of glucose, tolbutamide, glucocorticoid, growth hormone and glucagon. Diabetes 1973; 22(9): 687-93.
[http://dx.doi.org/10.2337/diab.22.9.687] [PMID: 4125576]

[48]   Rabinovitch A, Quigley C, Rechler MM. Growth hormone stimulates islet B-cell replication in neonatal rat pancreatic monolayer cultures. Diabetes 1983; 32(4): 307-12.
[http://dx.doi.org/10.2337/diab.32.4.307] [PMID: 6339303]

[49]   Nielsen JH, Linde S, Welinder BS, Billestrup N, Madsen OD. Growth hormone is a growth factor for the differentiated pancreatic β-cell. Mol Endocrinol 1989; 3(1): 165-73.
[http://dx.doi.org/10.1210/mend-3-1-165] [PMID: 2644530]

[50]   Møldrup A, Billestrup N, Nielsen JH. Rat insulinoma cells express both a 115-kDa growth hormone receptor and a 95-kDa prolactin receptor structurally related to the hepatic receptors. J Biol Chem 1990; 265(15): 8686-90.
[PMID: 1692834]

[51]   Brelje TC, Sorenson RL. Nutrient and hormonal regulation of the threshold of glucose-stimulated insulin secretion in isolated rat pancreases. Endocrinology 1988; 123(3): 1582-90.
[http://dx.doi.org/10.1210/endo-123-3-1582] [PMID: 3042373]

[52]   Sorenson RL, Brelje TC, Hegre OD, Marshall S, Anaya P, Sheridan JD. Prolactin (*in vitro*) decreases the glucose stimulation threshold, enhances insulin secretion, and increases dye coupling among islet B cells. Endocrinology 1987; 121(4): 1447-53.
[http://dx.doi.org/10.1210/endo-121-4-1447] [PMID: 3308438]

[53]   Weinhaus AJ, Stout LE, Bhagroo NV, Brelje TC, Sorenson RL. Regulation of glucokinase in pancreatic islets by prolactin: a mechanism for increasing glucose-stimulated insulin secretion during pregnancy. J Endocrinol 2007; 193(3): 367-81.

[http://dx.doi.org/10.1677/JOE-07-0043] [PMID: 17535875]

[54]   Weinhaus AJ, Stout LE, Sorenson RL. Glucokinase, hexokinase, glucose transporter 2, and glucose metabolism in islets during pregnancy and prolactin-treated islets *in vitro*: mechanisms for long term up-regulation of islets. Endocrinology 1996; 137(5): 1640-9.
        [PMID: 8612496]

[55]   Parsons JA, Brelje TC, Sorenson RL. Adaptation of islets of Langerhans to pregnancy: increased islet cell proliferation and insulin secretion correlates with the onset of placental lactogen secretion. Endocrinology 1992; 130(3): 1459-66.
        [PMID: 1537300]

[56]   Brelje TC, Scharp DW, Lacy PE, *et al.* Effect of homologous placental lactogens, prolactins, and growth hormones on islet B-cell division and insulin secretion in rat, mouse, and human islets: implication for placental lactogen regulation of islet function during pregnancy. Endocrinology 1993; 132(2): 879-87.
        [PMID: 8425500]

[57]   Kawai M, Kishi K. *In vitro* studies of the stimulation of insulin secretion and B-cell proliferation by rat placental lactogen-II during pregnancy in rats. J Reprod Fertil 1997; 109(1): 145-52.
        [http://dx.doi.org/10.1530/jrf.0.1090145] [PMID: 9068426]

[58]   Brelje TC, Sorenson RL. The physiological roles of prolactin, growth hormone and placental lactogen in the regulation of islet β cell proliferation. In: Sarvetnick N, Ed. Pancreatic Growth and Regeneration. Austin, USA: Karger Landes Systems 1997; pp. 1-30.

[59]   Labriola L, Montor WR, Krogh K, *et al.* Beneficial effects of prolactin and laminin on human pancreatic islet-cell cultures. Mol Cell Endocrinol 2007; 263(1-2): 120-33.
        [http://dx.doi.org/10.1016/j.mce.2006.09.011] [PMID: 17081683]

[60]   Parnaud G, Bosco D, Berney T, *et al.* Proliferation of sorted human and rat β cells. Diabetologia 2008; 51(1): 91-100.
        [http://dx.doi.org/10.1007/s00125-007-0855-1] [PMID: 17994216]

[61]   Kulkarni RN, Mizrachi EB, Ocana AG, Stewart AF. Human β-cell proliferation and intracellular signaling: driving in the dark without a road map. Diabetes 2012; 61(9): 2205-13.
        [http://dx.doi.org/10.2337/db12-0018] [PMID: 22751699]

[62]   Vasavada RC, Garcia-Ocaña A, Zawalich WS, *et al.* Targeted expression of placental lactogen in the β cells of transgenic mice results in β cell proliferation, islet mass augmentation, and hypoglycemia. J Biol Chem 2000; 275(20): 15399-406.
        [http://dx.doi.org/10.1074/jbc.275.20.15399] [PMID: 10809775]

[63]   Freemark M, Avril I, Fleenor D, *et al.* Targeted deletion of the PRL receptor: effects on islet development, insulin production, and glucose tolerance. Endocrinology 2002; 143(4): 1378-85.
        [http://dx.doi.org/10.1210/endo.143.4.8722] [PMID: 11897695]

[64]   Nielsen JH, Gittes G. Expression of growth hormone and prolactin receptors in the developing mouse pancreas. J Cell Biochem 1992; 16F (Suppl.): 88.

[65]   Huang C, Snider F, Cross JC. Prolactin receptor is required for normal glucose homeostasis and modulation of β-cell mass during pregnancy. Endocrinology 2009; 150(4): 1618-26.

[http://dx.doi.org/10.1210/en.2008-1003] [PMID: 19036882]

[66]   Parsons JA, Bartke A, Sorenson RL. Number and size of islets of Langerhans in pregnant, human growth hormone-expressing transgenic, and pituitary dwarf mice: effect of lactogenic hormones. Endocrinology 1995; 136(5): 2013-21.
[PMID: 7720649]

[67]   Liu JL, Coschigano KT, Robertson K, *et al.* Disruption of growth hormone receptor gene causes diminished pancreatic islet size and increased insulin sensitivity in mice. Am J Physiol Endocrinol Metab 2004; 287(3): E405-13.
[http://dx.doi.org/10.1152/ajpendo.00423.2003] [PMID: 15138153]

[68]   Wu Y, Liu C, Sun H, *et al.* Growth hormone receptor regulates β cell hyperplasia and glucose-stimulated insulin secretion in obese mice. J Clin Invest 2011; 121(6): 2422-6.
[http://dx.doi.org/10.1172/JCI45027] [PMID: 21555853]

[69]   Drynda R, Peters CJ, Jones PM, Bowe JE. The role of non-placental signals in the adaptation of islets to pregnancy. Horm Metab Res 2015; 47(1): 64-71.
[PMID: 25506682]

[70]   Møldrup A, Petersen ED, Nielsen JH. Effects of sex and pregnancy hormones on growth hormone and prolactin receptor gene expression in insulin-producing cells. Endocrinology 1993; 133(3): 1165-72.
[PMID: 8365359]

[71]   Polak M, Scharfmann R, Ban E, Haour F, Postel-Vinay MC, Czernichow P. Demonstration of lactogenic receptors in rat endocrine pancreases by quantitative autoradiography. Diabetes 1990; 39(9): 1045-9.
[http://dx.doi.org/10.2337/diab.39.9.1045] [PMID: 2166698]

[72]   Tesone M, Oliveira-Filho RM, Charreau EH. Prolactin binding in rat Langerhans islets. J Recept Res 1980; 1(3): 355-72.
[http://dx.doi.org/10.3109/10799898009038787] [PMID: 6271957]

[73]   Sorenson RL, Stout LE. Prolactin receptors and JAK2 in islets of Langerhans: an immunohistochemical analysis. Endocrinology 1995; 136(9): 4092-8.
[PMID: 7649117]

[74]   Brelje TC, Sorenson RL. Role of prolactin *versus* growth hormone on islet B-cell proliferation *in vitro*: implications for pregnancy. Endocrinology 1991; 128(1): 45-57.
[http://dx.doi.org/10.1210/endo-128-1-45] [PMID: 1986937]

[75]   Billestrup N, Nielsen JH. The stimulatory effect of growth hormone, prolactin, and placental lactogen on β-cell proliferation is not mediated by insulin-like growth factor-I. Endocrinology 1991; 129(2): 883-8.
[http://dx.doi.org/10.1210/endo-129-2-883] [PMID: 1677331]

[76]   Swenne I, Hill DJ, Strain AJ, Milner RD. Effects of human placental lactogen and growth hormone on the production of insulin and somatomedin C/insulin-like growth factor I by human fetal pancreas in tissue culture. J Endocrinol 1987; 113(2): 297-303.
[http://dx.doi.org/10.1677/joe.0.1130297] [PMID: 3295105]

[77]   Hill DJ, Frazer A, Swenne I, Wirdnam PK, Milner RD. Somatomedin-C in human fetal pancreas.

Cellular localization and release during organ culture. Diabetes 1987; 36(4): 465-71.
[http://dx.doi.org/10.2337/diab.36.4.465] [PMID: 3545947]

[78]   Scharfmann R, Corvol M, Czernichow P. Characterization of insulinlike growth factor I produced by fetal rat pancreatic islets. Diabetes 1989; 38(6): 686-90.
[http://dx.doi.org/10.2337/diab.38.6.686] [PMID: 2656337]

[79]   Van Schravendijk CF, Foriers A, Van den Brande JL, Pipeleers DG. Evidence for the presence of type I insulin-like growth factor receptors on rat pancreatic α and β cells. Endocrinology 1987; 121(5): 1784-8.
[http://dx.doi.org/10.1210/endo-121-5-1784] [PMID: 2959469]

[80]   Kulkarni RN, Holzenberger M, Shih DQ, *et al.* beta-cell-specific deletion of the Igf1 receptor leads to hyperinsulinemia and glucose intolerance but does not alter β-cell mass. Nat Genet 2002; 31(1): 111-5.
[PMID: 11923875]

[81]   Ma F, Wei Z, Shi C, *et al.* Signaling cross talk between growth hormone (GH) and insulin-like growth factor-I (IGF-I) in pancreatic islet β-cells. Mol Endocrinol 2011; 25(12): 2119-33.
[http://dx.doi.org/10.1210/me.2011-1052] [PMID: 22034225]

[82]   Nielsen JH. Growth and function of the pancreatic β cell *in vitro*: effects of glucose, hormones and serum factors on mouse, rat and human pancreatic islets in organ culture. Acta Endocrinol 1985; 108(Suppl. 266): 1-39.
[PMID: 3922192]

[83]   Dalgaard LT, Billestrup N, Nielsen JH. STAT5 activity in pancreatic β cells. Expert Rev Endocrinol Metab 2008; 3: 423-39.
[http://dx.doi.org/10.1586/17446651.3.4.423]

[84]   Friedrichsen BN, Galsgaard ED, Nielsen JH, Møldrup A. Growth hormone- and prolactin-induced proliferation of insulinoma cells, INS-1, depends on activation of STAT5 (signal transducer and activator of transcription 5). Mol Endocrinol 2001; 15(1): 136-48.
[http://dx.doi.org/10.1210/mend.15.1.0576] [PMID: 11145745]

[85]   Friedrichsen BN, Richter HE, Hansen JA, *et al.* Signal transducer and activator of transcription 5 activation is sufficient to drive transcriptional induction of cyclin D2 gene and proliferation of rat pancreatic β cells. Mol Endocrinol 2003; 17(5): 954-58.
[http://dx.doi.org/http://dx.doi.org/10.1210/me.2002-0356] [PMID: 12586844]

[86]   Brelje TC, Stout LE, Bhagroo NV, Sorenson RL. Distinctive roles for prolactin and growth hormone in the activation of signal transducer and activator of transcription 5 in pancreatic islets of langerhans. Endocrinology 2004; 145(9): 4162-75.
[http://dx.doi.org/10.1210/en.2004-0201] [PMID: 15142985]

[87]   Galsgaard ED, Nielsen JH, Møldrup A. Regulation of prolactin receptor (PRLR) gene expression in insulin-producing cells. Prolactin and growth hormone activate one of the rat prlr gene promoters *via* STAT5a and STAT5b. J Biol Chem 1999; 274(26): 18686-92.
[http://dx.doi.org/10.1074/jbc.274.26.18686] [PMID: 10373481]

[88]   Hansen JA, Lindberg K, Hilton DJ, Nielsen JH, Billestrup N. Mechanism of inhibition of growth hormone receptor signaling by suppressor of cytokine signaling proteins. Mol Endocrinol 1999; 13(11): 1832-43.

[http://dx.doi.org/10.1210/mend.13.11.0368] [PMID: 10551777]

[89]   Rønn SG, Hansen JA, Lindberg K, Karlsen AE, Billestrup N. The effect of suppressor of cytokine signaling 3 on GH signaling in β-cells. Mol Endocrinol 2002; 16(9): 2124-34.
[http://dx.doi.org/10.1210/me.2002-0082] [PMID: 12198248]

[90]   Lindberg K, Rønn SG, Tornehave D, *et al*. Regulation of pancreatic β-cell mass and proliferation by SOCS-3. J Mol Endocrinol 2005; 35(2): 231-43.
[http://dx.doi.org/10.1677/jme.1.01840] [PMID: 16216905]

[91]   Karlsen AE, Rønn SG, Lindberg K, *et al*. Suppressor of cytokine signaling 3 (SOCS-3) protects β - cells against interleukin-1β - and interferon-gamma -mediated toxicity. Proc Natl Acad Sci USA 2001; 98(21): 12191-6.
[http://dx.doi.org/10.1073/pnas.211445998] [PMID: 11593036]

[92]   Jensen J, Galsgaard ED, Karlsen AE, Lee YC, Nielsen JH. STAT5 activation by human GH protects insulin-producing cells against interleukin-1beta, interferon-gamma and tumour necrosis factor-alph- -induced apoptosis independent of nitric oxide production. J Endocrinol 2005; 187(1): 25-36.
[http://dx.doi.org/10.1677/joe.1.06086] [PMID: 16214938]

[93]   Fujinaka Y, Takane K, Yamashita H, Vasavada RC. Lactogens promote β cell survival through JAK2/STAT5 activation and Bcl-XL upregulation. J Biol Chem 2007; 282(42): 30707-17.
[http://dx.doi.org/10.1074/jbc.M702607200] [PMID: 17728251]

[94]   Jackerott M, Møldrup A, Thams P, *et al*. STAT5 activity in pancreatic β-cells influences the severity of diabetes in animal models of type 1 and 2 diabetes. Diabetes 2006; 55(10): 2705-12.
[http://dx.doi.org/10.2337/db06-0244] [PMID: 17003334]

[95]   Zhang F, Sjöholm A, Zhang Q. Growth hormone signaling in pancreatic β-cellscalcium handling regulated by growth hormone. Mol Cell Endocrinol 2009; 297(1-2): 50-7.
[http://dx.doi.org/10.1016/j.mce.2008.06.001] [PMID: 18602447]

[96]   Billestrup N, Møldrup A, Serup P, Mathews LS, Norstedt G, Nielsen JH. Introduction of exogenous growth hormone receptors augments growth hormone-responsive insulin biosynthesis in rat insulinoma cells. Proc Natl Acad Sci USA 1990; 87(18): 7210-4.
[http://dx.doi.org/10.1073/pnas.87.18.7210] [PMID: 2205855]

[97]   Bordin S, Amaral ME, Anhê GF, *et al*. Prolactin-modulated gene expression profiles in pancreatic islets from adult female rats. Mol Cell Endocrinol 2004; 220(1-2): 41-50.
[http://dx.doi.org/10.1016/j.mce.2004.04.001] [PMID: 15196698]

[98]   Porges U. Ueber Hypoglykämie bei Morbus Addison sowie bei nebennieren losen Hunden. Zeit f klin Med 1909; 69: 341.

[99]   Haist RE. Effects of steroids on the pancreas. In: Dorfmann RI, Ed. Methods in Hormone Research 4, part B. New York: Academic Press 1965; pp. 193-233.

[100]   Hellerström C. Growth pattern of pancreatic islets in animals. In: Volk BW, Wellmann KF, Eds. The diabetic pancreas. London: Ballière Tidall 1977; pp. 61-97.
[http://dx.doi.org/10.1007/978-1-4684-2325-9_3]

[101]   Brunstedt J, Nielsen JH. Direct long-term effect of hydrocortisone on insulin and glucagon release from mouse pancreatic islets in tissue culture. Acta Endocrinol 1981; 96(4): 498-504.

[PMID: 7010864]

[102] Nielsen JH. Hormonal regulation of growth and function of insulin-producing cells in culture. In: Fischer G, Wieser RJ, Eds. Hormonally defined media A tool in Cell Biology. Berlin: Springer Verlag 1983; pp. 264-74.
[http://dx.doi.org/10.1007/978-3-642-69290-1_31]

[103] Weinhaus AJ, Bhagroo NV, Brelje TC, Sorenson RL. Dexamethasone counteracts the effect of prolactin on islet function: implications for islet regulation in late pregnancy. Endocrinology 2000; 141(4): 1384-93.
[PMID: 10746642]

[104] Sutter-Dub MT. Effects of pregnancy and progesterone and/or oestradiol on the insulin secretion and pancreatic insulin content in the perfused rat pancreas. Diabetes Metab 1979; 5(1): 47-56.
[PMID: 446833]

[105] Howell SL, Tyhurst M, Green IC. Direct effects of progesterone on rat islets of Langerhans *in vivo* and in tissue culture. Diabetologia 1977; 13(6): 579-83.
[http://dx.doi.org/10.1007/BF01236310] [PMID: 338405]

[106] Nielsen JH, Nielsen v, Pedersen LM, Deckert T. Effects of pregnancy hormones on pancreatic islets in organ culture. Acta Endocrinol 1986; 111(3): 336-41.
[PMID: 3515818]

[107] Sorenson RL, Brelje TC, Roth C. Effects of steroid and lactogenic hormones on islets of Langerhans: a new hypothesis for the role of pregnancy steroids in the adaptation of islets to pregnancy. Endocrinology 1993; 133(5): 2227-34.
[PMID: 8404674]

[108] Nieuwenhuizen AG, Schuiling GA, Hilbrands LG, Bisschop EM, Koiter TR. Proliferation of pancreatic islet-cells in cyclic and pregnant rats after treatment with progesterone. Hormo- Metabol Res 1998; 30: 649-55.
[http://dx.doi.org/10.1055/s-2007-978952]

[109] Nieuwenhuizen AG, Schuiling GA, Liem SM, Moes H, Koiter TR, Uilenbroek JT. Progesterone stimulates pancreatic cell proliferation *in vivo*. Eur J Endocrinol 1999; 140: 256-63.

[110] Rodriguez RR. Influence of oestrogens and androgens on the production and prevention of diabetes. In: Leibel BS, Wrenshall GA, Eds. On the Nature and Treatment of Diabetes. Amsterdam: Excerpta Medica Foundation 1965; pp. 288-307.

[111] Houssay BA, Foglia VG, Rodriguez RR. Production or prevention of some types of experimental diabetes by oestrogens or corticosteroids. Acta Endocrinol 1954; 17(1-4): 146-64.
[PMID: 13227825]

[112] El Seifi S, Green IC, Perrin D. Insulin release and steroid-hormone binding in isolated islets of langerhans in the rat: effects of ovariectomy. J Endocrinol 1981; 90(1): 59-67.
[http://dx.doi.org/10.1677/joe.0.0900059] [PMID: 7021742]

[113] Green IC, El Seifi S, Perrin D, Howell SL. Cell replication in the islets of langerhans of adult rats: effects of pregnancy, ovariectomy and treatment with steroid hormones. J Endocrinol 1981; 88(2): 219-24.

[http://dx.doi.org/10.1677/joe.0.0880219] [PMID: 7009774]

[114] Nadal A, Rovira JM, Laribi O, *et al.* Rapid insulinotropic effect of 17beta-estradiol *via* a plasma membrane receptor. FASEB J 1998; 12: 1341-8.

[115] Nadal A, Alonso-Magdalena P, Soriano S, Quesada I, Ropero AB. The pancreatic β-cell as a target of estrogens and xenoestrogens: Implications for blood glucose homeostasis and diabetes. Mol Cell Endocrinol 2009; 304(1-2): 63-8.
[http://dx.doi.org/10.1016/j.mce.2009.02.016] [PMID: 19433249]

[116] Le May C, Chu K, Hu M, *et al.* Estrogens protect pancreatic β-cells from apoptosis and prevent insulin-deficient diabetes mellitus in mice. Proc Natl Acad Sci USA 2006; 103(24): 9232-7.
[http://dx.doi.org/10.1073/pnas.0602956103] [PMID: 16754860]

[117] Jacovetti C, Abderrahmani A, Parnaud G, *et al.* MicroRNAs contribute to compensatory β cell expansion during pregnancy and obesity. J Clin Invest 2012; 122(10): 3541-51.
[http://dx.doi.org/10.1172/JCI64151] [PMID: 22996663]

[118] Hakonen E, Ustinov J, Mathijs I, *et al.* Epidermal growth factor (EGF)-receptor signalling is needed for murine β cell mass expansion in response to high-fat diet and pregnancy but not after pancreatic duct ligation. Diabetologia 2011; 54(7): 1735-43.
[http://dx.doi.org/10.1007/s00125-011-2153-1] [PMID: 21509441]

[119] Hakonen E, Ustinov J, Palgi J, Miettinen PJ, Otonkoski T. EGFR signaling promotes β-cell proliferation and survivin expression during pregnancy. PLoS One 2014; 9(4): e93651.
[http://dx.doi.org/10.1371/journal.pone.0093651] [PMID: 24695557]

[120] Oda K, Matsuoka Y, Funahashi A, Kitano H. A comprehensive pathway map of epidermal growth factor receptor signaling. Mol Syst Biol 2005; 1(2005): 0010.
[http://dx.doi.org/10.1038/msb4100014]

[121] Rachdi L, Balcazar N, Osorio-Duque F, *et al.* Disruption of Tsc2 in pancreatic β cells induces β cell mass expansion and improved glucose tolerance in a TORC1-dependent manner. Proc Natl Acad Sci USA 2008; 105(27): 9250-5.
[http://dx.doi.org/10.1073/pnas.0803047105] [PMID: 18587048]

[122] Zahr E, Molano RD, Pileggi A, *et al.* Rapamycin impairs β-cell proliferation *in vivo*. Transplant Proc 2008; 40(2): 436-7.
[http://dx.doi.org/10.1016/j.transproceed.2008.02.011] [PMID: 18374093]

[123] Rieck S, White P, Schug J, *et al.* The transcriptional response of the islet to pregnancy in mice. Mol Endocrinol 2009; 23(10): 1702-12.
[http://dx.doi.org/10.1210/me.2009-0144] [PMID: 19574445]

[124] Scharfmann R, Tazi A, Polak M, Kanaka C, Czernichow P. Expression of functional nerve growth factor receptors in pancreatic β-cell lines and fetal rat islets in primary culture. Diabetes 1993; 42(12): 1829-36.
[http://dx.doi.org/10.2337/diab.42.12.1829] [PMID: 8243829]

[125] Demirci C, Ernst S, Alvarez-Perez JC, *et al.* Loss of HGF/c-Met signaling in pancreatic β-cells leads to incomplete maternal β-cell adaptation and gestational diabetes mellitus. Diabetes 2012; 61(5): 1143-52.

[http://dx.doi.org/10.2337/db11-1154] [PMID: 22427375]

[126] Araújo TG, Oliveira AG, Carvalho BM, *et al*. Hepatocyte growth factor plays a key role in insulin resistance-associated compensatory mechanisms. Endocrinology 2012; 153(12): 5760-9.
[http://dx.doi.org/10.1210/en.2012-1496] [PMID: 23024263]

[127] Accornero P, Miretti S, Starvaggi Cucuzza L, Martignani E, Baratta M. Epidermal growth factor and hepatocyte growth factor cooperate to enhance cell proliferation, scatter, and invasion in murine mammary epithelial cells. J Mol Endocrinol 2010; 44(2): 115-25.
[http://dx.doi.org/10.1677/JME-09-0035] [PMID: 19850646]

[128] Johansson M, Mattsson G, Andersson A, Jansson L, Carlsson PO. Islet endothelial cells and pancreatic β-cell proliferation: studies *in vitro* and during pregnancy in adult rats. Endocrinology 2006; 147(5): 2315-24.
[http://dx.doi.org/10.1210/en.2005-0997] [PMID: 16439446]

[129] Tamm I, Wang Y, Sausville E, *et al*. IAP-family protein survivin inhibits caspase activity and apoptosis induced by Fas (CD95), Bax, caspases, and anticancer drugs. Cancer Res 1998; 58(23): 5315-20.
[PMID: 9850056]

[130] Kanwar JR, Kamalapuram SK, Kanwar RK. Targeting survivin in cancer: the cell-signalling perspective. Drug Discov Today 2011; 16(11-12): 485-94.
[http://dx.doi.org/10.1016/j.drudis.2011.04.001] [PMID: 21511051]

[131] Jiang Y, Nishimura W, Devor-Henneman D, *et al*. Postnatal expansion of the pancreatic β-cell mass is dependent on survivin. Diabetes 2008; 57(10): 2718-27.
[http://dx.doi.org/10.2337/db08-0170] [PMID: 18599523]

[132] Xu Y, Wang X, Gao L, *et al*. Prolactin-stimulated survivin induction is required for β cell mass expansion during pregnancy in mice. Diabetologia 2015; 58(9): 2064-73.
[http://dx.doi.org/10.1007/s00125-015-3670-0] [PMID: 26099856]

[133] Wang H, Gambosova K, Cooper ZA, *et al*. EGF regulates survivin stability through the Raf-1/ERK pathway in insulin-secreting pancreatic β-cells. BMC Mol Biol 2010; 11: 66.
[http://dx.doi.org/10.1186/1471-2199-11-66] [PMID: 20807437]

[134] Layden BT, Durai V, Newman MV, *et al*. Regulation of pancreatic islet gene expression in mouse islets by pregnancy. J Endocrinol 2010; 207(3): 265-79.
[http://dx.doi.org/10.1677/JOE-10-0298] [PMID: 20847227]

[135] Kim H, Toyofuku Y, Lynn FC, *et al*. Serotonin regulates pancreatic β cell mass during pregnancy. Nat Med 2010; 16(7): 804-8.
[http://dx.doi.org/10.1038/nm.2173] [PMID: 20581837]

[136] Gleason CE, Ning Y, Cominski TP, *et al*. Role of insulin-like growth factor-binding protein 5 (IGFBP5) in organismal and pancreatic β-cell growth. Mol Endocrinol 2010; 24(1): 178-92.
[http://dx.doi.org/10.1210/me.2009-0167] [PMID: 19897600]

[137] Miyakoshi N, Richman C, Kasukawa Y, Linkhart TA, Baylink DJ, Mohan S. Evidence that IGF-binding protein-5 functions as a growth factor. J Clin Invest 2001; 107(1): 73-81.
[http://dx.doi.org/10.1172/JCI10459] [PMID: 11134182]

[138] Johnson SK, Dennis RA, Barone GW, Lamps LW, Haun RS. Differential expression of insulin-like growth factor binding protein-5 in pancreatic adenocarcinomas: identification using DNA microarray. Mol Carcinog 2006; 45(11): 814-27.
[http://dx.doi.org/10.1002/mc.20203] [PMID: 16865675]

[139] Firth SM, Baxter RC. Cellular actions of the insulin-like growth factor binding proteins. Endocr Rev 2002; 23(6): 824-54.
[http://dx.doi.org/10.1210/er.2001-0033] [PMID: 12466191]

[140] Schneider MR, Wolf E, Hoeflich A, Lahm H. IGF-binding protein-5: flexible player in the IGF system and effector on its own. J Endocrinol 2002; 172(3): 423-40.
[http://dx.doi.org/10.1677/joe.0.1720423] [PMID: 11874691]

[141] Ning Y, Hoang B, Schuller AG, *et al.* Delayed mammary gland involution in mice with mutation of the insulin-like growth factor binding protein 5 gene. Endocrinology 2007; 148(5): 2138-47.
[http://dx.doi.org/10.1210/en.2006-0041] [PMID: 17255210]

[142] Friedrichsen BN, Neubauer N, Lee YC, *et al.* Stimulation of pancreatic β-cell replication by incretins involves transcriptional induction of cyclin D1 *via* multiple signalling pathways. J Endocrinol 2006; 188(3): 481-92.
[http://dx.doi.org/10.1677/joe.1.06160] [PMID: 16522728]

[143] Buteau J, Foisy S, Joly E, Prentki M. Glucagon-like peptide 1 induces pancreatic β-cell proliferation *via* transactivation of the epidermal growth factor receptor. Diabetes 2003; 52(1): 124-32.
[http://dx.doi.org/10.2337/diabetes.52.1.124] [PMID: 12502502]

[144] Moffett RC, Vasu S, Thorens B, Drucker DJ, Flatt PR. Incretin receptor null mice reveal key role of GLP-1 but not GIP in pancreatic β cell adaptation to pregnancy. PLoS One 2014; 9(6): e96863.
[http://dx.doi.org/10.1371/journal.pone.0096863] [PMID: 24927416]

[145] Josefsen K, Lee YC, Thams P, Efendic S, Nielsen JH. AKAP 18 alpha and gamma have opposing effects on insulin release in INS-1E cells. FEBS Lett 2010; 584(1): 81-5.
[http://dx.doi.org/10.1016/j.febslet.2009.10.086] [PMID: 19896945]

[146] Nieuwenhuizen AG, Schuiling GA, Moes H, Koiter TR. Role of increased insulin demand in the adaptation of the endocrine pancreas to pregnancy. Acta Physiol Scand 1997; 159(4): 303-12.
[http://dx.doi.org/10.1046/j.1365-201X.1997.d01-1872.x] [PMID: 9146751]

[147] Nieuwenhuizen AG, Schuiling GA, Seijsener AF, Moes H, Koiter TR. Effects of food restriction on glucose tolerance, insulin secretion, and islet-cell proliferation in pregnant rats. Physiol Behav 1999; 65(4-5): 671-7.
[http://dx.doi.org/10.1016/S0031-9384(98)00203-0] [PMID: 10073466]

[148] Marynissen G, Aerts L, Van Assche FA. The endocrine pancreas during pregnancy and lactation in the rat. J Dev Physiol 1983; 5(6): 373-81.
[PMID: 6361113]

[149] Porat S, Weinberg-Corem N, Tornovsky-Babaey S, *et al.* Control of pancreatic β cell regeneration by glucose metabolism. Cell Metab 2011; 13(4): 440-9.
[http://dx.doi.org/10.1016/j.cmet.2011.02.012] [PMID: 21459328]

[150] Nakamura A, Togashi Y, Orime K, *et al.* Control of β cell function and proliferation in mice

stimulated by small-molecule glucokinase activator under various conditions. Diabetologia 2012; 55(6): 1745-54.
[http://dx.doi.org/10.1007/s00125-012-2521-5] [PMID: 22456697]

[151] Kassem S, Bhandari S, Rodríguez-Bada P, *et al.* Large islets, β-cell proliferation, and a glucokinase mutation. N Engl J Med 2010; 362(14): 1348-50.
[http://dx.doi.org/10.1056/NEJMc0909845] [PMID: 20375417]

[152] Brelje TC, Bhagroo NV, Stout LE, Sorenson RL. Beneficial effects of lipids and prolactin on insulin secretion and β-cell proliferation: a role for lipids in the adaptation of islets to pregnancy. J Endocrinol 2008; 197(2): 265-76.
[http://dx.doi.org/10.1677/JOE-07-0657] [PMID: 18434356]

[153] Abouna S, Old RW, Pelengaris S, *et al.* Non-β-cell progenitors of β-cells in pregnant mice. Organogenesis 2010; 6(2): 125-33.
[http://dx.doi.org/10.4161/org.6.2.10374] [PMID: 20885859]

[154] Rukstalis JM, Habener JF. Neurogenin3: a master regulator of pancreatic islet differentiation and regeneration. Islets 2009; 1(3): 177-84.
[http://dx.doi.org/10.4161/isl.1.3.9877] [PMID: 21099270]

[155] Søstrup B. Molecular mechanisms involved in β cell expansion during pregnancy. Denmark: University of Copenhagen 2013.

[156] Søstrup B, Gaarn LW, Nalla A, Billestrup N, Nielsen JH. Co-ordinated regulation of neurogenin-3 expression in the maternal and fetal pancreas during pregnancy. Acta Obstet Gynecol Scand 2014; 93(11): 1190-7.
[http://dx.doi.org/10.1111/aogs.12495] [PMID: 25179808]

[157] Zhao X. Increase of β cell mass by β cell replication, but not neogenesis, in the maternal pancreas in mice. Endocr J 2014; 61(6): 623-8.
[http://dx.doi.org/10.1507/endocrj.EJ14-0040] [PMID: 24748457]

[158] Xiao X, Chen Z, Shiota C, *et al.* No evidence for β cell neogenesis in murine adult pancreas. J Clin Invest 2013; 123(5): 2207-17.
[http://dx.doi.org/10.1172/JCI66323] [PMID: 23619362]

[159] Zhang H, Zhang J, Pope CF, *et al.* Gestational diabetes mellitus resulting from impaired β-cell compensation in the absence of FoxM1, a novel downstream effector of placental lactogen. Diabetes 2010; 59(1): 143-52.
[http://dx.doi.org/10.2337/db09-0050] [PMID: 19833884]

[160] Zhang H, Ackermann AM, Gusarova GA, *et al.* The FoxM1 transcription factor is required to maintain pancreatic β-cell mass. Mol Endocrinol 2006; 20(8): 1853-66.
[http://dx.doi.org/10.1210/me.2006-0056] [PMID: 16556734]

[161] Wang IC, Chen YJ, Hughes D, *et al.* Forkhead box M1 regulates the transcriptional network of genes essential for mitotic progression and genes encoding the SCF (Skp2-Cks1) ubiquitin ligase. Mol Cell Biol 2005; 25(24): 10875-94.
[http://dx.doi.org/10.1128/MCB.25.24.10875-10894.2005] [PMID: 16314512]

[162] Gupta RK, Gao N, Gorski RK, *et al.* Expansion of adult β-cell mass in response to increased metabolic

demand is dependent on HNF-4alpha. Genes Dev 2007; 21(7): 756-69.
[http://dx.doi.org/10.1101/gad.1535507] [PMID: 17403778]

[163] Gupta S, Davis RJ. MAP kinase binds to the NH2-terminal activation domain of c-Myc. FEBS Lett 1994; 353(3): 281-5.
[http://dx.doi.org/10.1016/0014-5793(94)01052-8] [PMID: 7957875]

[164] Horn S, Kirkegaard JS, Hoelper S, *et al.* Research Resource: A dual proteomic approach identifies regulated islet proteins during β-Cell mass expansion *in vivo.* Mol Endocrinol 2016; 30(1): 133-43.
[http://dx.doi.org/10.1210/me.2015-1208] [PMID: 26649805]

[165] Laybutt DR, Weir GC, Kaneto H, *et al.* Overexpression of c-Myc in β-cells of transgenic mice causes proliferation and apoptosis, downregulation of insulin gene expression, and diabetes. Diabetes 2002; 51(6): 1793-804.
[http://dx.doi.org/10.2337/diabetes.51.6.1793] [PMID: 12031967]

[166] Rieck S, Kaestner KH. Expansion of β-cell mass in response to pregnancy. Trends Endocrinol Metab 2010; 21(3): 151-8.
[http://dx.doi.org/10.1016/j.tem.2009.11.001] [PMID: 20015659]

[167] Ritzel RA, Butler PC. Replication increases β-cell vulnerability to human islet amyloid polypeptide-induced apoptosis. Diabetes 2003; 52(7): 1701-8.
[http://dx.doi.org/10.2337/diabetes.52.7.1701] [PMID: 12829636]

[168] Karnik SK, Chen H, McLean GW, *et al.* Menin controls growth of pancreatic β-cells in pregnant mice and promotes gestational diabetes mellitus. Science 2007; 318(5851): 806-9.
[http://dx.doi.org/10.1126/science.1146812] [PMID: 17975067]

[169] Yi P, Park JS, Melton DA. Betatrophin: a hormone that controls pancreatic β cell proliferation. Cell 2013; 153(4): 747-58.
[http://dx.doi.org/10.1016/j.cell.2013.04.008] [PMID: 23623304]

[170] Kondegowda NG, Fenutria R, Pollack IR, *et al.* Osteoprotegerin and denosumab stimulate human β cell proliferation through inhibition of the receptor activator of NF-κB ligand pathway. Cell Metab 2015; 22(1): 77-85.
[http://dx.doi.org/10.1016/j.cmet.2015.05.021] [PMID: 26094891]

[171] Hong JS, Santolaya-Forgas J, Romero R, *et al.* Maternal plasma osteoprotegerin concentration in normal pregnancy. Am J Obstet Gynecol 2005; 193(3 Pt 2): 1011-5.
[http://dx.doi.org/10.1016/j.ajog.2005.06.051] [PMID: 16157103]

[172] Mallette FA, Moiseeva O, Calabrese V, Mao B, Gaumont-Leclerc MF, Ferbeyre G. Transcriptome analysis and tumor suppressor requirements of STAT5-induced senescence. Ann N Y Acad Sci 2010; 1197: 142-51.
[http://dx.doi.org/10.1111/j.1749-6632.2010.05192.x] [PMID: 20536843]

[173] Kutlu B, Burdick D, Baxter D, *et al.* Detailed transcriptome atlas of the pancreatic β cell. BMC Med Genomics 2009; 2: 3.
[http://dx.doi.org/10.1186/1755-8794-2-3] [PMID: 19146692]

[174] Schraenen A, Lemaire K, de Faudeur G, *et al.* Placental lactogens induce serotonin biosynthesis in a subset of mouse β cells during pregnancy. Diabetologia 2010; 53(12): 2589-99.

[http://dx.doi.org/10.1007/s00125-010-1913-7] [PMID: 20938637]

[175]   Iida H, Ogihara T, Min MK, *et al.* Expression mechanism of tryptophan hydroxylase 1 in mouse islets during pregnancy. J Mol Endocrinol 2015; 55(1): 41-53.
[PMID: 26136513]

[176]   Wang G, Rajpurohit SK, Delaspre F, *et al.* First quantitative high-throughput screen in zebrafish identifies novel pathways for increasing pancreatic β-cell mass. eLife 2015; 4: 4.
[http://dx.doi.org/10.7554/eLife.08261] [PMID: 26218223]

[177]   Tsuji N, Ninov N, Delawary M, *et al.* Whole organism high content screening identifies stimulators of pancreatic β-cell proliferation. PLoS One 2014; 9(8): e104112.
[http://dx.doi.org/10.1371/journal.pone.0104112] [PMID: 25117518]

[178]   Lesurtel M, Graf R, Aleil B, *et al.* Platelet-derived serotonin mediates liver regeneration. Science 2006; 312(5770): 104-7.
[http://dx.doi.org/10.1126/science.1123842] [PMID: 16601191]

[179]   Ohara-Imaizumi M, Kim H, Yoshida M, *et al.* Serotonin regulates glucose-stimulated insulin secretion from pancreatic β cells during pregnancy. Proc Natl Acad Sci USA 2013; 110(48): 19420-5.
[http://dx.doi.org/10.1073/pnas.1310953110] [PMID: 24218571]

[180]   Rycyzyn MA, Reilly SC, OMalley K, Clevenger CV. Role of cyclophilin B in prolactin signal transduction and nuclear retrotranslocation. Mol Endocrinol 2000; 14(8): 1175-86.
[http://dx.doi.org/10.1210/mend.14.8.0508] [PMID: 10935542]

[181]   Kim J, Choi TG, Ding Y, *et al.* Overexpressed cyclophilin B suppresses apoptosis associated with ROS and Ca^{2+} homeostasis after ER stress. J Cell Sci 2008; 121(Pt 21): 3636-48.
[http://dx.doi.org/10.1242/jcs.028654] [PMID: 18946027]

[182]   Schimmack S, Taylor A, Lawrence B, *et al.* Stathmin in pancreatic neuroendocrine neoplasms: a marker of proliferation and PI3K signaling. Tumour Biol 2015; 36(1): 399-408.
[http://dx.doi.org/10.1007/s13277-014-2629-y] [PMID: 25266798]

[183]   Shin JE, Geisler S, DiAntonio A. Dynamic regulation of SCG10 in regenerating axons after injury. Exp Neurol 2014; 252: 1-11.
[http://dx.doi.org/10.1016/j.expneurol.2013.11.007] [PMID: 24246279]

[184]   Carlsson C, Tornehave D, Lindberg K, *et al.* Growth hormone and prolactin stimulate the expression of rat preadipocyte factor-1/delta-like protein in pancreatic islets: molecular cloning and expression pattern during development and growth of the endocrine pancreas. Endocrinology 1997; 138(9): 3940-8.
[PMID: 9275085]

[185]   Friedrichsen BN, Carlsson C, *et al.* Expression, biosynthesis and release of preadipocyte factor-1/delta-like protein/fetal antigen-1 in pancreatic β-cells: possible physiological implications. J Endocrinol 2003; 176(2): 257-66.
[http://dx.doi.org/http://dx.doi.org/10.1677/joe.0.1760257] [PMID: 12553874]

[186]   Jackerott M, Lee YC, Møllgård K, *et al.* Trefoil factors are expressed in human and rat endocrine pancreas: differential regulation by growth hormone. Endocrinology 2006; 147(12): 5752-9.
[http://dx.doi.org/10.1210/en.2006-0601] [PMID: 16973727]

[187] Fueger PT, Schisler JC, Lu D, *et al.* Trefoil factor 3 stimulates human and rodent pancreatic islet β-cell replication with retention of function. Mol Endocrinol 2008; 22(5): 1251-9.
[http://dx.doi.org/10.1210/me.2007-0500] [PMID: 18258687]

[188] Samson MH, Vestergaard EM, Milman N, Poulsen SS, Nexø E. Circulating serum trefoil factors increase dramatically during pregnancy. Scand J Clin Lab Invest 2008; 68(5): 369-74.
[http://dx.doi.org/10.1080/00365510701767862] [PMID: 19172695]

[189] Nalla A, Ringholm L, Søstrup B, *et al.* Implications for the offspring of circulating factors involved in β cell adaptation in pregnancy. Acta Obstet Gynecol Scand 2014; 93(11): 1181-9.
[http://dx.doi.org/10.1111/aogs.12505] [PMID: 25223212]

[190] Kalis M, Kumar R, Janciauskiene S, Salehi A, Cilio CM. α 1-antitrypsin enhances insulin secretion and prevents cytokine-mediated apoptosis in pancreatic β-cells. Islets 2010; 2(3): 185-9.
[http://dx.doi.org/10.4161/isl.2.3.11654] [PMID: 21099312]

[191] Bernichtein S, Kayser C, Dillner K, *et al.* Development of pure prolactin receptor antagonists. J Biol Chem 2003; 278(38): 35988-99.
[http://dx.doi.org/10.1074/jbc.M305687200] [PMID: 12824168]

[192] Ben-Harousb A, Yogev Y, Hod M. Epidemiology of gestational diabetes mellitus. In: Hod M, Janovic L, Di Renzo GC, De Leiva A, Langer O, Eds. Textbook of Diabetes and Pregnancy. London, United Kingdom: Informa UK Ltd. 2008; pp. 118-31.
[http://dx.doi.org/10.3109/9781439802007.015]

[193] Devlieger R, Casteels K, Van Assche FA. Reduced adaptation of the pancreatic B cells during pregnancy is the major causal factor for gestational diabetes: current knowledge and metabolic effects on the offspring. Acta Obstet Gynecol Scand 2008; 87(12): 1266-70.
[http://dx.doi.org/10.1080/00016340802443863] [PMID: 18846453]

[194] Nielsen LR, Rehfeld JF, Pedersen-Bjergaard U, Damm P, Mathiesen ER. Pregnancy-induced rise in serum C-peptide concentrations in women with type 1 diabetes. Diabetes Care 2009; 32(6): 1052-7.
[http://dx.doi.org/10.2337/dc08-1832] [PMID: 19244092]

[195] Nalla A. Adaptation of pancreatic β cells to pregnancy: Role of circulating factors from pregnant women without and with autoimmune diabetes on β cell growth and survival *in vitro*. Denmark: University of Copenhagen 2012.

[196] Atwater I, Gondos B, DiBartolomeo R, Bazaes R, Jovanovic L. Pregnancy hormones prevent diabetes and reduce lymphocytic infiltration of islets in the NOD mouse. Ann Clin Lab Sci 2002; 32(1): 87-92.
[PMID: 11848623]

[197] Holstad M, Sandler S. Prolactin protects against diabetes induced by multiple low doses of streptozotocin in mice. J Endocrinol 1999; 163(2): 229-34.
[http://dx.doi.org/10.1677/joe.0.1630229] [PMID: 10556772]

[198] Lau J, Börjesson A, Holstad M, Sandler S. Prolactin regulation of the expression of TNF-alpha, IFN-gamma and IL-10 by splenocytes in murine multiple low dose streptozotocin diabetes. Immunol Lett 2006; 102(1): 25-30.
[http://dx.doi.org/10.1016/j.imlet.2005.06.006] [PMID: 16054232]

[199] Bellamy L, Casas JP, Hingorani AD, Williams D. Type 2 diabetes mellitus after gestationsal diabetes:

a systematic review and meta-analysis. Lancet 2009; 373: 1773-9.
[http://dx.doi.org/http://dx.doi.org/10.1016/s0140-6736(09)60731-5]

[200] Kühl C. Etiology and pathogenesis of gestational diabetes. Diabetes Care 1998; 21 (Suppl. 2): B19-26.
[PMID: 9704223]

[201] Nielsen JH, Haase TN, Jaksch C, *et al.* Impact of fetal and neonatal environment on β cell function and development of diabetes. Acta Obstet Gynecol Scand 2014; 93(11): 1109-22.
[http://dx.doi.org/10.1111/aogs.12504] [PMID: 25225114]

[202] Prentice KJ, Luu L, Allister EM, *et al.* The furan fatty acid metabolite CMPF is elevated in diabetes and induces β cell dysfunction. Cell Metab 2014; 19(4): 653-66.
[http://dx.doi.org/10.1016/j.cmet.2014.03.008] [PMID: 24703697]

[203] Talchai C, Xuan S, Lin HV, Sussel L, Accili D. Pancreatic β cell dedifferentiation as a mechanism of diabetic β cell failure. Cell 2012; 150(6): 1223-34.
[http://dx.doi.org/10.1016/j.cell.2012.07.029] [PMID: 22980982]

[204] Scaglia L, Smith FE, Bonner-Weir S. Apoptosis contributes to the involution of β cell mass in the post partum rat pancreas. Endocrinology 1995; 136(12): 5461-8.
[PMID: 7588296]

[205] Gunderson EP, Crites Y, Chiang V, *et al.* Influence of breastfeeding during the postpartum oral glucose tolerance test on plasma glucose and insulin. Obstet Gynecol 2012; 120(1): 136-43.
[http://dx.doi.org/10.1097/AOG.0b013e31825b993d] [PMID: 22914402]

CHAPTER 8

β-Cells from Embryonic and Adult Stem Cells and Progenitors

Christine A. Beamish*

Department of Surgery, Islet Transplantation Laboratory, The Methodist Hospital Research Institute, Houston Texas, USA

Abstract: Diabetes is a chronic autoimmune disease, causing the destruction of the insulin-producing β-cells of the pancreatic islet and leading to glycemic dysregulation. Exogenous insulin administration and glucose testing moderately rectifies hyperglycemia, but does not provide adequate fine tuning necessary for complete prevention of hypoglycemia acutely, nor micro- and macro-vascular complications in the long-term. Islet transplants have shown great promise for this dynamic glucose regulation, but a shortage of cadaveric-sourced cells, and lifelong immune suppression requirements vastly restrict this technique from being widely available to patients with the disease. Therefore alternative sources of insulin-producing cells are needed. In this chapter, the role of stem cell biology in the current context of diabetes therapy is discussed, including an assessment of human embryonic and human induced pluripotent stem cells for the restoration of β-cell mass. Additionally, the existence of putative resident stem cells, and possible fluidity in lineage fate determination within endocrine pancreas- related cell types is examined.

Keywords: β-cell, Diabetes, Pancreas, Plasticity, Progenitor cell, Regeneration, Stem cell.

INTRODUCTION

Type 1 diabetes mellitus is an autoimmune disease of the endocrine pancreas, involving an interplay of immune, genetic, and environmentally-mediated factors

*Corresponding author Christine A. Beamish:** Department of Surgery, Islet Transplantation Laboratory, The Methodist Hospital Research Institute, Houston Texas, USA; Tel: 713-363-9193; Fax: 713-441-3240; E-mail: cbeamish@houstonmethodist.org

David J. Hill (Ed.)

[1 - 3]. While the overarching goal of diabetes therapy would ostensibly restrict the autoimmune destruction of insulin-producing β-cells and retain existing β-cell mass in susceptible individuals, the primary impediment to this approach is that the disease only presents with overt symptoms of hyperglycemia after β-cell mass is lost below a clinical threshold (~10% remaining), and long after disease initiation [4]. Furthermore, while research advances actively target a variety of parameters including auto-antibody presence, diabetes-correlative genetics such as the presence of the HLA alleles DR3-DQ2/DR4-DQ8 [3], and viral exposure [2], <20% of patients have known risk factors prior to diagnosis (such as first-degree relatives with diabetes) [5], thus limiting the ability to predict diabetes incidence in all but few patients [6, 7]. Therefore, resolving the immune destruction is only part of the solution, and β-cell replacement strategies are imperative.

Since its advent in 1989 by Lacy, Ricordi, Scharp and associates [8, 9], islet transplants have improved drastically as a treatment for specific T1D patients. Much of the improvement came in the year 2000 with the Edmonton Protocol pioneered by Shapiro and colleagues [10], allowing exogenous insulin independence effected by a higher islet mass from two donors, and the use of steroid-free immunosuppression *via* sirolimus, tacrolimus, and daclizumab, an IL2-receptor antagonist antibody [11]. Grafts now last up to 5 years [12, 13]. In addition to freedom from injections, benefits from islet transplantation include significant decreases in micro- and macro-vascular complication rates, decreases in severe hypoglycemic episodes, increased circulating C-peptide titers, and improvement in HbA$_{1c}$ [13, 14]. Importantly, these benefits are noted even if the graft fails. The obstacle to this therapy being offered more widely is primarily due to lack of islets, as well as risks associated with life-long immunosuppression for allogeneic grafts [13]. This shortage of islets is multi-factorial: many available pancreata do not meet minimum criteria for transplant, including donor age and metabolic profile; islets may be damaged irreparably during isolation; and tragically, the majority of islets die in the first days after transplant [15], primarily resulting from an instant blood mediated immune reaction (IBMIR) [16]. Some research has focused on mitigating the factors responsible for this process, such as modulating platelet-monocyte interactions [17] or by blocking complement activation [18]. The remaining islet death results from the loss of blood supply, causing hypoxia, ischemia-reperfusion injury [19], and amyloid deposition as time progresses. Islet vascularization is a critical determinant of cell survival in the long-term [20]. Cost of the procedure is a final regional consideration, resulting from islet transplants being designated as "experimental therapy" in the United States, and hence only (and rarely) covered by private insurance or research funding, although this is not the case in other countries, such as Canada [21].

Alternate sources of insulin-producing β-cells would alleviate the need for cadaveric-sourced human tissue with their associated sequelae, and significantly alter morbidity and mortality outcomes in the future. The potential role of stem cell biology will be examined to this end, including assessments of both exogenous and endogenous stem and progenitor cell populations and their applications to human medicine.

GENERATION OF β-CELLS FROM STEM CELL SOURCE

The *de novo* generation of insulin-producing β-cells has long been a goal for diabetes therapy. Given the expansive data generated from lineage tracing experiments, it is now possible to ascertain the sequence of steps necessary in the developmental biology of the β-cell, from primitive gut endoderm through to functional, glucose-responsive, β-cell [22]. The generation of β-cells from stem cell source has generally co-opted these developmental procedures.

A stem cell may be defined by two functional properties, namely an unlimited ability for self-renewal, and the capacity to generate multiple cell types (*e.g.* multi-potentiality). This may be, but is not limited to, a response to injury or other stimulus. A progenitor cell, alternatively, demonstrates some restriction in self-renewal capability and which is usually uni-potent [23, 24]. The distinction between these two cell states is often fluid and vaguely indeterminate in the literature, but which may yield important and definitive outcomes for medicine. Similarly, subdivisions in definition exist for the somatic, or tissue-specific, stem cells *vs.* embryonic stem cells, namely that the latter demonstrates the capacity to generate progeny across all primary germ layers. However, precedents exist to suggest that this is not always rigidly obeyed, as has been demonstrated by hematopoietic and neural stem cells forming tissues from germ layers other than mesoderm- and ectoderm-lineage, respectively [25 - 28].

Human embryonic stem (hES) cells are a natural choice for the generation of β-cells and have been used extensively for this purpose, now widely available from commercial cell lines. The β-like-cells produced by early experiments had limited function and were generally glucose-unresponsive [29 - 31]. Enhancements to these procedures yielded moderately improved results, engineered by the use of multi-stage, stepwise procedures which recapitulate endocrine pancreas maturation, and utilize signaling by various molecules including those from the TGF-β superfamily members Activin-A/Nodal [32, 33], as well as Wnt [33 - 36]. The differentiation of hES cells to definitive endoderm was shown to require SOX17 [37] and FOXA2/HNF3β [38], as well as other small chemical effectors [39]. However, fully glucose responsive β-cells generated by hES cell differentiation remained frustratingly elusive, instead producing cells that more closely resembled human fetal (immature) β-cells rather than functional, mature ones [40], and which required long-term transplantation *in vivo* to effect significant mitigation of hyperglycemia [41].

Important progress in the generation of functional, glucose responsive β-cells was recently accomplished by the Kieffer laboratory, by finer manipulation of *in vitro* conditions, enabling a more efficient transition to pancreatic endoderm, and, by extension endocrine cells. This was achieved by inducing higher expression of PDX1+NKX6.1+ pancreatic progenitors and thus higher insulin content at the completion of the experiment [42], as well as fewer poly-hormonal (lineage-unspecified) cells [43]. The protocol modifications included the sequential addition of various small molecules, including Alk5i/Noggin [44], thyroid hormone, and a gamma secretase inhibitor [42].

However, these cells still required transplantation for 3-4 months before cell maturation and glucose amelioration was accomplished. The Melton laboratory methodically addressed this problem using >150 combinations of >70 compounds, arriving at a protocol utilizing 11 different factors and ≥ 5 weeks culture time [45]. Some notable differences in protocol include an extension in culture time with keratinocyte growth factor (KGF), the hedgehog inhibitor SANT1, and a reduced concentration of retinoic acid (RA). These improvements were shown to generate β-cells capable of dynamic and repeated response to glucose, and importantly, cells which did not require transplantation for maturation [45]. A general schematic of the steps required to generate a functional β-cell from human pluripotent cell source is shown in Fig. (**1**).

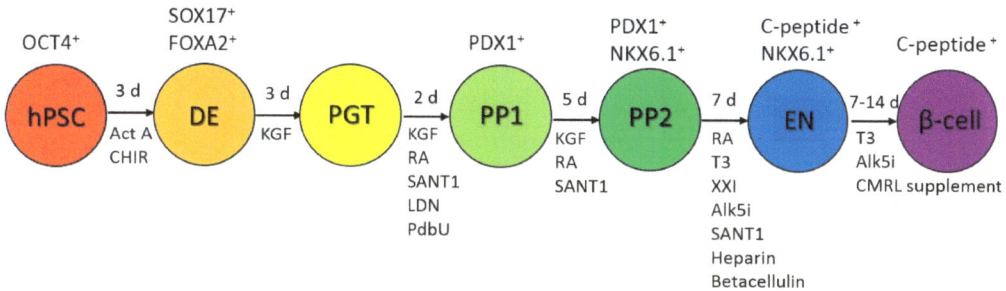

Fig. (1). Generation of β-cells in vitro from stem cell source. Schematic of directed differentiation from hPSC (human pluripotent stem cells, including induced pluripotent and embryonic stem cells) into glucose responsive, insulin+ cells. Multi-step addition of signalling molecules and growth factors yields cells with progressively defined identity and function, from definitive endoderm (DE) to primitive gut tube (PGT), early pancreatic progenitor cells (PP1), later pancreatic progenitor cells (PP2), endocrine cells (EN), and finally, functional β-cells. Act A, Activin A; CHIR, CHIR99021, a GSK3α/β inhibitor and WNT activator; KGF, keratinocyte growth factor; RA, retinoic acid; SANT1, sonic hedgehog pathway antagonist; LDN, LDN193189, a BMP type 1 receptor inhibitor; PdbU, Phorbol 12,13-dibutyrate, a protein kinase C activator; Alk5i, Alk5 receptor inhibitor II; T3, triiodothyronine, thyroid hormone; XXI, γ-secretase inhibitor; βcellulin, EGF family member.

The scope of this protocol is shown in its scalability and reproducibility for human transplant procedures, with one patient requiring cells from 1-2 flasks, *versus* the current requirement of 1-3 human donor pancreata per recipient [45]. This scalability is crucial for clinical translation, as are strategies to enhance graft survival such as implantation devices including cell differentiation cassettes (Viacyte™) [46], silicon-channeled Nanoglands [47], gelatin foam sponges [41], or hydrogel spheres [48]. These bio-engineered structures are then often coated in alginate or other similar material [49] which aim to limit recipient immune-mediated foreign body responses (FBR). Proof-of-principle for this technique was recently shown by the Anderson laboratory in conjunction with Melton and colleagues, in which human stem-cell derived β-cells (SC- β-cells) were encapsulated with 1.5 mm triazole thiomorpholine dioxide (TMTD) alginate and transplanted into immune-competent hyperglycemic mice, which demonstrated a robust insulin response and, notably, survival for 174 days with minimal fibrosis.

Alternatively, β-like-cells have been similarly generated by the directed reprogramming of induced pluripotent stem (iPS) cells, which may be obtained from the patient directly and thus theoretically bypass the need for some immunosuppression regimes. Furthermore, iPS cells avoid the potential socio-political ramifications that arise from the manipulation of human embryonic tissue. IPS cells are generated from fibroblasts obtained from skin biopsies, blood cells, or cell lines, subsequently transduced with lentiviral vectors containing the transcription factors OCT4, SOX2, KLF4, and C-MYC, which generate cellular colonies *in vitro* [35, 45, 50 - 52]. Differentiation of iPS cells is then achieved using the same protocols as used by hES cell culture, *via* sequential induction to definitive endoderm, through to early pancreatic progenitor cells, and into insulin-producing cells, with seemingly equivalent results [42, 45, 53].

REGENERATION OF THE ENDOCRINE PANCREAS

Data from genetic manipulation and directed programming of hES and iPS cells show unprecedented promise and there is much hope for diabetes therapy within this realm. However, these practices are not without their own inherent risks, such as the potential for tumor formation [51], nor is it known the short or long-term functionality of these *de novo* β-cells in a human context, including kinetics and dynamic responses to glucose [54]. Moreover, there are numerous examples of β-cell regeneration *in vivo*, which may be exploited for diabetes therapy prior to the introduction of foreign material. In overly simplistic terms, if autoimmune destruction can be appreciably controlled, there may be the potential to allow the pancreas to heal itself, at least in those patients with some retained β-cell function early after diagnosis. While β-cell proliferation in adulthood has been shown to be minimal [55, 56], expansion of β-cell mass has been demonstrated in certain circumstances of stress or growth, and covered in detail in previous chapters, such as during pregnancy, obesity, and aging. Self-duplication is widely accepted to be the primary mode for β-cell replacement [57], however there are increasing examples of regeneration not accounted for by replication of existing β-cells alone. Is endogenous stem or progenitor cell activation possible within the endocrine pancreas?

Generation of Insulin-Expressing Cells *In Vitro*

Cellular identity is broadly determined by the interplay of genetic (and epigenetic) and environmental factors which control gene expression in a cellular system [58]. Therefore, instructive changes to either/both the environment or the genes present may effect a change in cell fate. Since all pancreatic cell types arise from a common progenitor cell type, *in vitro* differentiation is a conceivable approach. Dedifferentiation can be defined as the loss of mature, defining and functional characteristics from a partially or terminally differentiated cell type [58], whereas trans-differentiation is broadly defined as a change from one differentiated phenotype to another, involving morphological and functional phenotypic markers [59]. These operational terminologies are utilized to describe the differentiation, dedifferentiation, or trans-differentiation of pancreatic cells towards or away from the genetic, morphological, and functional properties of mature β-cells. Importantly, cellular trans-differentiation can be induced by a multitude of factors, such as direct reprogramming induced by transcription factor over-expression [60 - 62], or alternatively by modified progression through dedifferentiation to a multipotent progenitor-like stage, and subsequent redifferentiation. These cellular fate (re)specifications are adapted from Puri *et al.* [58] in Fig. (**2**), and show that terminal cellular differentiation in the endocrine pancreas is not entirely unidirectional.

Multiple reports have previously demonstrated that human fetal pancreas cells could be induced to expand and express insulin [63 - 69]. This was accomplished by *in vitro* techniques, such as extracellular matrix selection [67, 70, 71] and specific combinations of growth factors including hepatocyte growth factor (HGF) [72], fibroblast growth factor-2 (FGF-2), KGF, and insulin-like growth factor – II (IGF-II) [69, 73], activin, βcellulin, exendin-4, and HGF [74], basic FGF (bFGF) and leukemia inhibitory factor (LIF) [75], epidermal growth factor (EGF) and cholera toxin [71, 76], or insulin/ transferring/selenium (ITS) [77], and generally the use of serum-free media. Many groups contended that the resultant cells, with extensive proliferation of fibroblast-like cells, had arisen from β-cells, due to the expression of insulin [78, 79]. However, the insulin generated from *in vitro* culture was much lower than would be found *in vivo*, and there remains the possibility of progenitor cell utilization.

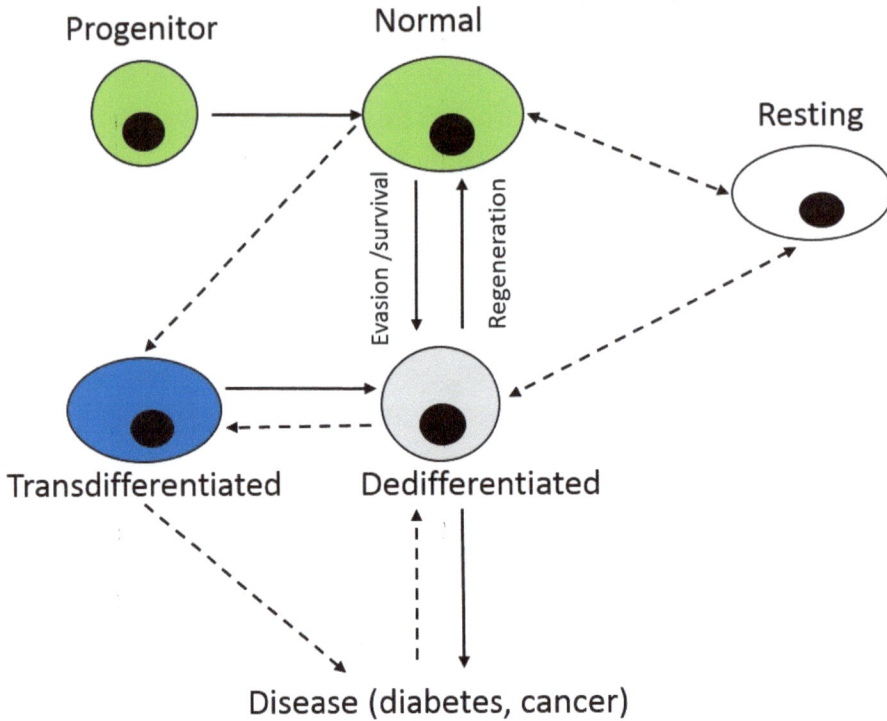

Fig. (2). Fate decisions made by the β-cell during health and disease. Illustration of differentiation potential of cells, resulting from either stress/disease, or by genetic and/or environmental manipulation. Note that "transdifferentiation" can occur by direct means, or by prior "dedifferentiation". Regeneration is hypothesized to occur by a return to normal from the dedifferentiated state, whereas the reverse is proposed to occur during severe β-cell loss or stress (Evasion/survival). These fate re-specifications may contribute to disease states, including diabetes and cancer; however, determination of the reversibility of these processes remains a goal for medicine.

The advent of lineage tracing technology allowed β-cell fate to be followed during *in vitro* culture conditions, and demonstrated that the majority of mouse β-cells were, in fact, eliminated from culture [74, 75, 77, 80, 81]. However, some studies were able to demonstrate the transition of a small subset (<5%) of adult mouse β-cells [80, 81] or adult human β-cells [82] to non-insulin-expressing states, showing β-cell dedifferentiation potential. It remains to be determined why some β-cells are capable of differentiation whereas the majority of them are not. Interestingly, it emerged that human β-cells potentially retain a higher plasticity potential than do mouse β-cells [80, 83], although the mechanisms behind this have not been elucidated.

Resident Stem/Progenitor Cells Within the Endocrine Pancreas

The search for stem/progenitor cells present within the endocrine pancreas itself has yielded little information, as proto-typical "stem" cells have not been readily identified within the pancreas as can be found in other organs, such as intestine [24]. However, recent paradigm shifts have focused the search towards the endocrine pancreas itself as the source of regeneration. Szabat *et al.* traced human and mouse β-cells with a dual Ins/Pdx1 reporter system, and reported that a subpopulation of adult β-cells from both species exhibited negative/low expression of insulin while maintaining expression of Pdx1 (Pdx1$^+$Ins^{low}cells) in 15-25% of cells. These could differentiate into Pdx1$^+$Ins$^+$ cells without cell division, implying differentiation [84]. Furthermore, these Pdx1$^+$Ins^{low} cells displayed a progenitor phenotype, lacking expression of Glut2, Gck, and MafA, whilst being proliferative. Interestingly, these Pdx1$^+$Ins^{low} cells could be sub-divided into two groups: those that matured into insulin-expressing cells, and those that maintained a progenitor phenotype [84].

A separate study examined the impact of nutritional insult in early life on β-cell regeneration potential [85]. It emerged that there was no effect on Pdx1$^+$Ins$^-$ cell number when mice were protein-restricted during gestation, or treated with the β-cell toxin streptozotocin (STZ) independently. However, the cells were present in higher proportion in pancreas exposed to STZ and low-protein diet simultaneously, suggesting that the ability of progenitor cells to mature and differentiate to functioning β-cells was mitigated as a result of the *in utero* insults [85].

Data from Liu and colleagues directly contradicted work by Dor *et al.* [57]: lineage tracing of the β-cell using the same RIPCreER;Z/AP$^{+/+}$ transgenic mouse model resulted in an increase in HPAP$^+$ (reporter) β-cell labeling with age, as well as the presence of HPAP$^+$ insulin$^-$ cells after aging and STZ-mediated injury [86]. They found that HPAP$^+$Ins$^-$ cells demonstrated an immature phenotype, with increased expression of Pdx1 and MafB, and the lack of Nkx6.1 and Glut2. Furthermore, HPAP$^+$ Ins$^-$ cells were proliferative after STZ as demonstrated by expression of Ki67, and were 2-fold smaller than typical β-cells. The relative proportion of HPAP⁺Ins⁻ cells after injury was found to represent a small minority (0.1%) of all insulin⁺ (β-) cells. The authors concluded that these cells represented β-cell progenitors which, by default of expressing the β-cell reporter, must have exhibited active insulin expression at the time of Cre induction. Only after injury or aging, and hence after differentiation to "mature" β-cells, was insulin expressed, which accounted for the increase in HPAP⁺ labeled Ins⁺ cells with increased age [86].

The van der Kooy laboratory has demonstrated similarly provocative evidence for resident pancreatic stem cells. Seaberg *et al.* found that rare cells within mouse islets and ductal preparations could form clonal spheres with the capacity to generate neural and pancreatic exocrine and endocrine lineages [87]; these cells could then be further differentiated into functional β-cells which demonstrated glucose-sensitive insulin release. The group determined that these cells were not derived from embryonic stem cell origin, nor from neural crest, nor from mesoderm lineage, but instead from "multipotent precursor" (progenitor) cells found within the pancreas [87]. Smukler *et al.* then replicated the clonal sphere assay [88], revealing that these multipotent cells had derived from insulin-expressing cells. Specifically however, these multipotent cells were characterized to represent a unique subset of β-cells in that they demonstrated decreased levels of Nkx6.1 and Pdx1, increased levels of Ngn3, and importantly, lacked Glut2 [88]. These cells were present within adult mouse and human islets in a proportion of <1/5,000, and could be differentiated *in vitro* to multiple endocrine lineages (including α, β, δ, and γ cells), exocrine cells, and neural lineage cells including neurons, (neuro)glia, and oligodendrocytes [88]. Moreover, after differentiation to β-cells, these were fully capable of ameliorating hyperglycemia after transplant, demonstrating their utility as functional endocrine cells [88]. A subsequent report from the same laboratory has convincingly demonstrated a selective survival, proliferation, and differentiation potential of insulin⁺Glut2^{low} cells towards a β-cell phenotype after diabetes presence in both human and mouse [89], indicating an active role of progenitor cell utilization during disease.

New evidence supports insulin-positive but Glut2-low (Ins⁺Glut2^{LO}) progenitor cell presence within the mouse pancreas, and was found to be enriched in the neonatal mouse relative to the adult [90]. Importantly, these cells were present in much higher proportion outside of the islet, within small β-cell clusters containing less than 5 β-cells; these β-cell clusters are generally ignored. A subset of these glucose-unresponsive cells were capable of multi-lineage plasticity, which declined in number with age, concomitant with aggregation of β-cells into proto-typical islet structures.

Promiscuity in Cell Lineage Within the Endocrine Pancreas

The putative presence of resident progenitor cells within the endocrine pancreas raises multiple questions, including one central to cell biology: how predetermined, or set, is cell fate? The report by Smukler and colleagues, which challenged the findings by Dor *et al.* [57], found that HPAP[+] (reporter-expressing) cells tagged non-β-cells after a long chase period including endocrine and non-endocrine cells [88], and which was recently replicated [90]. Evidence that a minority of β-cells labeled non-β-cell progeny after pulse-chase experiments [86, 88, 90] indicates that cell fate may not be as immutable as assumed. Indeed, these data have been found previously, but attributed to a lack of transgene stringency ("Cre leakiness") [91] and dismissed as true cell fate switching. Early lineage tracing experiments by Herrera showed that adult α- and β-cells derive from independent lineages [92]; however, recent experiments by the same group have suggested that α-cells retain the ability to trans-differentiate to β-cells under conditions of extreme β-cell loss [93], and by forced Pdx1 expression [94]. This α- to- β lineage switch has been also shown by others, first after pancreatic duct ligation [95], and similarly after using caerulein as a chemical agent [96], and the β-cell toxin alloxan in both cases. After caerulein-induced trans-differentiation, the neogenic β-cells subsequently assumed a δ-cell fate [96]. Conversely, β-cells have been shown to trans-differentiate to α-cells during type 2 diabetes progression [97], further exaggerating the loss of β-cell mass. This result was postulated to represent stress-induced β-cell dedifferentiation, and hence failure [98]. In an injury model using partial pancreatectomy, β-cells lost insulin and gained pancreatic polypeptide (PP) expression, yielding a population of proliferative, Smad-7[+] cells, suggesting a similar dedifferentiation process [99]. Others have found that β-cell dedifferentiation could be evaluated by the relative presence of the peptide Urocortin 3 (Ucn3) [100, 101] which was used as a marker of maturation status. Furthermore, the dedifferentiated phenotype could be rescued by the addition of a small molecule (Alk5) which inhibited TGFβ receptor 1 [102]. Isolated human islets were shown to demonstrate an increase in β-cell proliferation and maturation at the expense of α-cell proportion when cultured with the decarboxylated form of the osteoblast-derived hormone, osteocalcin [103]. Recently, one group exploited the properties of cell fate indetermination, and induced trans-differentiation of α- to β-cell lineages using Activin signaling in an appropriate knockout mouse model [104], underscoring the potential role of directed reprogramming in the generation of insulin-producing cells.

These data collectively suggest that there exist redundant mechanisms within the pancreas to maintain the potential for insulin production, with dedifferentiation at times acting as a means to evade complete β-cell annihilation whilst attempting to sustain glucose homeostasis. Whether this survival technique is indeed useful in preserving β-cell mass, or is instead an erroneous fate decision which does not rectify the problem, remains to be determined. Regardless, these data suggest that the pancreas is, in fact, a dynamically regulating organ capable of incredible plasticity and regeneration. As we move into a new era of personalized medicine, how we generate new β-cells from stem cell source directly, or promote and protect the body's own regenerative capacity will be of utmost importance for the treatment of diabetes in the future.

CONFLICT OF INTEREST

The author confirms that author has no conflict of interest to declare for this publication.

ACKNOWLEDGEMENTS

Declared none.

REFERENCES

[1]   Szablewski L. Role of immune system in type 1 diabetes mellitus pathogenesis. Int Immunopharmacol 2014; 22(1): 182-91.
[http://dx.doi.org/10.1016/j.intimp.2014.06.033] [PMID: 24993340]

[2]   Craig ME, Nair S, Stein H, Rawlinson WD. Viruses and type 1 diabetes: a new look at an old story. Pediatr Diabetes 2013; 14(3): 149-58.

[http://dx.doi.org/10.1111/pedi.12033] [PMID: 23517503]

[3]   Erlich H, Valdes AM, Noble J, *et al.* HLA DR-DQ haplotypes and genotypes and type 1 diabetes risk: analysis of the type 1 diabetes genetics consortium families. Diabetes 2008; 57(4): 1084-92.
       [http://dx.doi.org/10.2337/db07-1331] [PMID: 18252895]

[4]   Lebastchi J, Herold KC. Immunologic and metabolic biomarkers of β-cell destruction in the diagnosis of type 1 diabetes. Cold Spring Harb Perspect Med 2012; 2(6): a007708.
       [http://dx.doi.org/10.1101/cshperspect.a007708] [PMID: 22675665]

[5]   Morran MP, Vonberg A, Khadra A, Pietropaolo M. Immunogenetics of type 1 diabetes mellitus. Mol Aspects Med 2015; 42: 42-60.
       [http://dx.doi.org/10.1016/j.mam.2014.12.004] [PMID: 25579746]

[6]   Simmons KM, Michels AW. Type 1 diabetes: A predictable disease. World J Diabetes 2015; 6(3): 380-90.
       [http://dx.doi.org/10.4239/wjd.v6.i3.380] [PMID: 25897349]

[7]   Ziegler AG, Nepom GT. Prediction and pathogenesis in type 1 diabetes. Immunity 2010; 32(4): 468-78.
       [http://dx.doi.org/10.1016/j.immuni.2010.03.018] [PMID: 20412757]

[8]   Ricordi C, Lacy PE, Scharp DW. Automated islet isolation from human pancreas. Diabetes 1989; 38 (Suppl. 1): 140-2.
       [http://dx.doi.org/10.2337/diab.38.1.S140] [PMID: 2642838]

[9]   Scharp DW, Lacy PE, Santiago JV, *et al.* Insulin independence after islet transplantation into type I diabetic patient. Diabetes 1990; 39(4): 515-8.
       [http://dx.doi.org/10.2337/diab.39.4.515] [PMID: 2108071]

[10]   Shapiro AM, Lakey JR, Ryan EA, *et al.* Islet transplantation in seven patients with type 1 diabetes mellitus using a glucocorticoid-free immunosuppressive regimen. N Engl J Med 2000; 343(4): 230-8.
       [http://dx.doi.org/10.1056/NEJM200007273430401] [PMID: 10911004]

[11]   Bruni A, Gala-Lopez B, Pepper AR, Abualhassan NS, Shapiro AJ. Islet cell transplantation for the treatment of type 1 diabetes: recent advances and future challenges. Diabetes Metab Syndr Obes 2014; 7: 211-23.
       [http://dx.doi.org/10.2147/DMSO.S50789] [PMID: 25018643]

[12]   Bellin MD, Barton FB, Heitman A, *et al.* Potent induction immunotherapy promotes long-term insulin independence after islet transplantation in type 1 diabetes. Am J Transplant 2012; 12(6): 1576-83.
       [http://dx.doi.org/10.1111/j.1600-6143.2011.03977.x] [PMID: 22494609]

[13]   Barton FB, Rickels MR, Alejandro R, *et al.* Improvement in outcomes of clinical islet transplantation: 19992010. Diabetes Care 2012; 35(7): 1436-45.
       [http://dx.doi.org/10.2337/dc12-0063] [PMID: 22723582]

[14]   Thompson DM, Meloche M, Ao Z, *et al.* Reduced progression of diabetic microvascular complications with islet cell transplantation compared with intensive medical therapy. Transplantation 2011; 91(3): 373-8.
       [http://dx.doi.org/10.1097/TP.0b013e31820437f3] [PMID: 21258272]

[15]   Liljebäck H, Grapensparr L, Olerud J, Carlsson PO. Extensive loss of islet mass beyond the first day

after intraportal human islet transplantation in a mouse model. Cell Transplant 2016; 25(3): 481-9.
[http://dx.doi.org/10.3727/096368915X688902] [PMID: 26264975]

[16]   Hårdstedt M, Lindblom S, Karlsson-Parra A, Nilsson B, Korsgren O. Characterization of innate immunity in an extended whole blood model of human islet allotransplantation. Cell Transplant 2016; 25(3): 503-15.
[http://dx.doi.org/10.3727/096368915X688461] [PMID: 26084381]

[17]   Kourtzelis I, Kotlabova K, Lim JH, *et al.* Developmental endothelial locus-1 modulates platelet-monocyte interactions and instant blood-mediated inflammatory reaction in islet transplantation. Thromb Haemost 2016; 115(4): 781-8.
[http://dx.doi.org/10.1160/TH15-05-0429] [PMID: 26676803]

[18]   Xiao F, Ma L, Zhao M, *et al.* APT070 (mirococept), a membrane-localizing C3 convertase inhibitor, attenuates early human islet allograft damage *in vitro* and *in vivo* in a humanized mouse model. Br J Pharmacol 2016; 173(3): 575-87.
[http://dx.doi.org/10.1111/bph.13388] [PMID: 26565566]

[19]   Chhabra P, Linden J, Lobo P, Okusa MD, Brayman KL. The immunosuppressive role of adenosine A2A receptors in ischemia reperfusion injury and islet transplantation. Curr Diabetes Rev 2012; 8(6): 419-33.
[http://dx.doi.org/10.2174/157339912803529878] [PMID: 22934547]

[20]   Brissova M, Aamodt K, Brahmachary P, *et al.* Islet microenvironment, modulated by vascular endothelial growth factor-A signaling, promotes β cell regeneration. Cell Metab 2014; 19(3): 498-511.
[http://dx.doi.org/10.1016/j.cmet.2014.02.001] [PMID: 24561261]

[21]   National Institute of Diabetes and Digestive and Kidney Disease. Pancreatic Islet Transplantation US: US Department of Health and Human Service 2013. Available at: http://www.niddk.nih.gov/ health-information/health-topics/Diabetes/pancreatic-islet-transplantation/Pages/index.aspx#2ref.

[22]   Pan FC, Wright C. Pancreas organogenesis: from bud to plexus to gland. Dev Dyn 2011; 240: 530-65.
[http://dx.doi.org/10.1002/dvdy.22584]

[23]   Seaberg RM, van der Kooy D. Stem and progenitor cells: the premature desertion of rigorous definitions. Trends Neurosci 2003; 26(3): 125-31.
[http://dx.doi.org/10.1016/S0166-2236(03)00031-6] [PMID: 12591214]

[24]   Potten CS, Loeffler M. Stem cells: attributes, cycles, spirals, pitfalls and uncertainties. Lessons for and from the crypt. Development 1990; 110(4): 1001-20.
[PMID: 2100251]

[25]   Clarke D, Johansson C, Wilbertz J, *et al.* Generalized potential of adult neural stem cells. Science 2000; 288: 1660-4.
[http://dx.doi.org/10.1126/science.288.5471.1660]

[26]   Brazelton TR, Rossi FM V, Keshet GI, Blau HM. From marrow to brain : expression of neuronal phenotypes in adult mice. Science 2000; 290: 1775-9.

[27]   Galli R, Borello U, Gritti A, *et al.* Skeletal myogenic potential of human and mouse neural stem cells. Nat Neurosci 2000; 3(10): 986-91.
[http://dx.doi.org/10.1038/79924] [PMID: 11017170]

[28]   Bjornson C, Rietze R, Reynolds B, Magli M, Vescovi A. Turning brain into blood : a hematopoietic fate adopted by adult neural stem cells *in vivo*. Science 2003; 442: 534-7.

[29]   Assady S, Maor G, Amit M, Itskovitz-Eldor J, Skorecki KL, Tzukerman M. Insulin production by human embryonic stem cells. Diabetes 2001; 50(8): 1691-7.
[http://dx.doi.org/10.2337/diabetes.50.8.1691] [PMID: 11473026]

[30]   Segev H, Fishman B, Ziskind A, Shulman M, Itskovitz-Eldor J. Differentiation of human embryonic stem cells into insulin-producing clusters. Stem Cells 2004; 22(3): 265-74.
[http://dx.doi.org/10.1634/stemcells.22-3-265] [PMID: 15153604]

[31]   Roche E, Sepulcre P, Reig JA, Santana A, Soria B. Ectodermal commitment of insulin-producing cells derived from mouse embryonic stem cells. FASEB J 2005; 19(10): 1341-3.
[PMID: 15928194]

[32]   Chen AE, Borowiak M, Sherwood RI, Kweudjeu A, Melton DA. Functional evaluation of ES cell-derived endodermal populations reveals differences between Nodal and Activin A-guided differentiation. Development 2013; 140(3): 675-86.
[http://dx.doi.org/10.1242/dev.085431] [PMID: 23293299]

[33]   DAmour KA, Bang AG, Eliazer S, *et al*. Production of pancreatic hormone-expressing endocrine cells from human embryonic stem cells. Nat Biotechnol 2006; 24(11): 1392-401.
[http://dx.doi.org/10.1038/nbt1259] [PMID: 17053790]

[34]   Nostro MC, Cheng X, Keller GM, Gadue P. Wnt, activin, and BMP signaling regulate distinct stages in the developmental pathway from embryonic stem cells to blood. Cell Stem Cell 2008; 2(1): 60-71.
[http://dx.doi.org/10.1016/j.stem.2007.10.011] [PMID: 18371422]

[35]   Nostro MC, Sarangi F, Ogawa S, *et al*. Stage-specific signaling through TGFβ family members and WNT regulates patterning and pancreatic specification of human pluripotent stem cells. Development 2011; 138(5): 861-71.
[http://dx.doi.org/10.1242/dev.055236] [PMID: 21270052]

[36]   Nostro MC, Keller G. Generation of β cells from human pluripotent stem cells: Potential for regenerative medicine. Semin Cell Dev Biol 2012; 23(6): 701-10.
[http://dx.doi.org/10.1016/j.semcdb.2012.06.010] [PMID: 22750147]

[37]   Kanai-Azuma M, Kanai Y, Gad JM, *et al*. Depletion of definitive gut endoderm in Sox17-null mutant mice. Development 2002; 129(10): 2367-79.
[PMID: 11973269]

[38]   Li Z, Gadue P, Chen K, *et al*. Foxa2 and H2A.Z mediate nucleosome depletion during embryonic stem cell differentiation. Cell 2012; 151(7): 1608-16.
[http://dx.doi.org/10.1016/j.cell.2012.11.018] [PMID: 23260146]

[39]   Borowiak M, Maehr R, Chen S, *et al*. Small molecules efficiently direct endodermal differentiation of mouse and human embryonic stem cells. Cell Stem Cell 2009; 4(4): 348-58.
[http://dx.doi.org/10.1016/j.stem.2009.01.014] [PMID: 19341624]

[40]   Hrvatin S, ODonnell CW, Deng F, *et al*. Differentiated human stem cells resemble fetal, not adult, β cells. Proc Natl Acad Sci USA 2014; 111(8): 3038-43.
[http://dx.doi.org/10.1073/pnas.1400709111] [PMID: 24516164]

[41] Kroon E, Martinson LA, Kadoya K, *et al.* Pancreatic endoderm derived from human embryonic stem cells generates glucose-responsive insulin-secreting cells *in vivo.* Nat Biotechnol 2008; 26(4): 443-52.
[http://dx.doi.org/10.1038/nbt1393] [PMID: 18288110]

[42] Rezania A, Bruin JE, Arora P, *et al.* Reversal of diabetes with insulin-producing cells derived *in vitro* from human pluripotent stem cells. Nat Biotechnol 2014; 32(11): 1121-33.
[http://dx.doi.org/10.1038/nbt.3033] [PMID: 25211370]

[43] Bruin JE, Erener S, Vela J, *et al.* Characterization of polyhormonal insulin-producing cells derived *in vitro* from human embryonic stem cells. Stem Cell Res (Amst) 2014; 12(1): 194-208.
[http://dx.doi.org/10.1016/j.scr.2013.10.003] [PMID: 24257076]

[44] Rezania A, Bruin JE, Riedel MJ, *et al.* Maturation of human embryonic stem cell-derived pancreatic progenitors into functional islets capable of treating pre-existing diabetes in mice. Diabetes 2012; 61(8): 2016-29.
[http://dx.doi.org/10.2337/db11-1711] [PMID: 22740171]

[45] Pagliuca FW, Millman JR, Gürtler M, *et al.* Generation of functional human pancreatic β cells *in vitro.* Cell 2014; 159(2): 428-39.
[http://dx.doi.org/10.1016/j.cell.2014.09.040] [PMID: 25303535]

[46] Schulz TC, Young HY, Agulnick AD, *et al.* A scalable system for production of functional pancreatic progenitors from human embryonic stem cells. PLoS One 2012; 7(5): e37004.
[http://dx.doi.org/10.1371/journal.pone.0037004] [PMID: 22623968]

[47] Sabek OM, Ferrati S, Fraga DW, *et al.* Characterization of a nanogland for the autotransplantation of human pancreatic islets. Lab Chip 2013; 13(18): 3675-88.
[http://dx.doi.org/10.1039/c3lc50601k] [PMID: 23884326]

[48] Veiseh O, Doloff JC, Ma M, *et al.* Size- and shape-dependent foreign body immune response to materials implanted in rodents and non-human primates. Nat Mater 2015; 14(6): 643-51.
[http://dx.doi.org/10.1038/nmat4290] [PMID: 25985456]

[49] Motté E, Szepessy E, Suenens K, *et al.* Composition and function of macroencapsulated human embryonic stem cell-derived implants: comparison with clinical human islet cell grafts. Am J Physiol Endocrinol Metab 2014; 307(9): E838-46.
[http://dx.doi.org/10.1152/ajpendo.00219.2014] [PMID: 25205822]

[50] Thatava T, Nelson TJ, Edukulla R, *et al.* Indolactam V/GLP-1-mediated differentiation of human iPS cells into glucose-responsive insulin-secreting progeny. Gene Ther 2011; 18(3): 283-93.
[http://dx.doi.org/10.1038/gt.2010.145] [PMID: 21048796]

[51] Thatava T, Kudva YC, Edukulla R, *et al.* Intrapatient variations in type 1 diabetes-specific iPS cell differentiation into insulin-producing cells. Mol Ther 2013; 21(1): 228-39.
[http://dx.doi.org/10.1038/mt.2012.245] [PMID: 23183535]

[52] Maehr R, Chen S, Snitow M, *et al.* Generation of pluripotent stem cells from patients with type 1 diabetes. Proc Natl Acad Sci USA 2009; 106(37): 15768-73.
[http://dx.doi.org/10.1073/pnas.0906894106] [PMID: 19720998]

[53] Yabe SG, Fukuda S, Takeda F, Nashiro K, Shimoda M, Okochi H. Efficient generation of functional pancreatic β cells from human iPS cells. J Diabetes 2016. [Epub ahead of print].

[http://dx.doi.org/10.1111/1753-0407.12400] [PMID: 27038181]

[54]   Kushner JA, MacDonald PE, Atkinson MA. Stem cells to insulin secreting cells: two steps forward and now a time to pause? Cell Stem Cell 2014; 15(5): 535-6.
[http://dx.doi.org/10.1016/j.stem.2014.10.012] [PMID: 25517460]

[55]   Teta M, Long SY, Wartschow LM, Rankin MM, Kushner JA. Very slow turnover of β-cells in aged adult mice. Diabetes 2005; 54(9): 2557-67.
[http://dx.doi.org/10.2337/diabetes.54.9.2557] [PMID: 16123343]

[56]   Meier JJ, Butler AE, Saisho Y, *et al.* β-cell replication is the primary mechanism subserving the postnatal expansion of β-cell mass in humans. Diabetes 2008; 57(6): 1584-94.
[http://dx.doi.org/10.2337/db07-1369] [PMID: 18334605]

[57]   Dor Y, Brown J, Martinez OI, Melton DA. Adult pancreatic β-cells are formed by self-duplication rather than stem-cell differentiation. Nature 2004; 429(6987): 41-6.
[http://dx.doi.org/10.1038/nature02520] [PMID: 15129273]

[58]   Puri S, Folias AE, Hebrok M. Plasticity and dedifferentiation within the pancreas: development, homeostasis, and disease. Cell Stem Cell 2014; 1-14.
[http://dx.doi.org/10.1016/j.stem.2014.11.001] [PMID: 25465113]

[59]   Okada T. Transdifferentiation in animal models: fact or artifact? Dev Growth Differ 1986; 28: 213-21.
[http://dx.doi.org/10.1111/j.1440-169X.1986.00213.x]

[60]   Yechoor V, Liu V, Espiritu C, *et al.* Neurogenin3 is sufficient for transdetermination of hepatic progenitor cells into neo-islets *in vivo* but not transdifferentiation of hepatocytes. Dev Cell 2009; 16(3): 358-73.
[http://dx.doi.org/10.1016/j.devcel.2009.01.012] [PMID: 19289082]

[61]   Zhou Q, Brown J, Kanarek A, Rajagopal J, Melton DA. *In vivo* reprogramming of adult pancreatic exocrine cells to β-cells. Nature 2008; 455(7213): 627-32.
[http://dx.doi.org/10.1038/nature07314] [PMID: 18754011]

[62]   Chen YJ, Finkbeiner SR, Weinblatt D, *et al.* De novo formation of insulin-producing neo-β cell islets from intestinal crypts. Cell Reports 2014; 6(6): 1046-58.
[http://dx.doi.org/10.1016/j.celrep.2014.02.013] [PMID: 24613355]

[63]   Beattie GM, Hayek A, Levine F. Growth and genetic modification of human β-cells and β-cell precursors. Genet Eng (N Y) 2000; 22: 99-120.
[PMID: 11501383]

[64]   Beattie GM, Leibowitz G, Lopez AD, Levine F, Hayek A. Protection from cell death in cultured human fetal pancreatic cells. Cell Transplant 2000; 9(3): 431-8.
[PMID: 10972342]

[65]   Demeterco C, Beattie GM, Dib SA, Lopez AD, Hayek A. A role for activin A and βcellulin in human fetal pancreatic cell differentiation and growth. J Clin Endocrinol Metab 2000; 85(10): 3892-7.
[PMID: 11061554]

[66]   Halvorsen TL, Beattie GM, Lopez AD, Hayek A, Levine F. Accelerated telomere shortening and senescence in human pancreatic islet cells stimulated to divide *in vitro*. J Endocrinol 2000; 166(1): 103-9.

[http://dx.doi.org/10.1677/joe.0.1660103] [PMID: 10856888]

[67]   Beattie GM, Rubin JS, Mally MI, Otonkoski T, Hayek A. Regulation of proliferation and differentiation of human fetal pancreatic islet cells by extracellular matrix, hepatocyte growth factor, and cell-cell contact. Diabetes 1996; 45(9): 1223-8.
       [http://dx.doi.org/10.2337/diab.45.9.1223] [PMID: 8772726]

[68]   Beattie GM, Itkin-Ansari P, Cirulli V, *et al.* Sustained proliferation of PDX-1+ cells derived from human islets. Diabetes 1999; 48(5): 1013-9.
       [http://dx.doi.org/10.2337/diabetes.48.5.1013] [PMID: 10331405]

[69]   Otonkoski T, Beattie GM, Rubin JS, Lopez AD, Baird A, Hayek A. Hepatocyte growth factor/scatter factor has insulinotropic activity in human fetal pancreatic cells. Diabetes 1994; 43(7): 947-53.
       [http://dx.doi.org/10.2337/diab.43.7.947] [PMID: 8013761]

[70]   Kerr-Conte J, Pattou F, Lecomte-Houcke M, *et al.* Ductal cyst formation in collagen-embedded adult human islet preparations. A means to the reproduction of nesidioblastosis *in vitro.* Diabetes 1996; 45(8): 1108-14.
       [http://dx.doi.org/10.2337/diab.45.8.1108] [PMID: 8690159]

[71]   Yuan S, Rosenberg L, Paraskevas S, Agapitos D, Duguid WP. Transdifferentiation of human islets to pancreatic ductal cells in collagen matrix culture. Differentiation 1996; 61(1): 67-75.
       [http://dx.doi.org/10.1046/j.1432-0436.1996.6110067.x] [PMID: 8921586]

[72]   Kayali AG, Flores LE, Lopez AD, *et al.* Limited capacity of human adult islets expanded *in vitro* to redifferentiate into insulin-producing β-cells. Diabetes 2007; 56(3): 703-8.
       [http://dx.doi.org/10.2337/db06-1545] [PMID: 17327439]

[73]   Movassat J, Beattie GM, Lopez AD, Portha B, Hayek A. Keratinocyte growth factor and β-cell differentiation in human fetal pancreatic endocrine precursor cells. Diabetologia 2003; 46(6): 822-9.
       [http://dx.doi.org/10.1007/s00125-003-1117-5] [PMID: 12802496]

[74]   Chase LG, Ulloa-Montoya F, Kidder BL, Verfaillie CM. Islet-derived fibroblast-like cells are not derived *via* epithelial-mesenchymal transition from Pdx-1 or insulin-positive cells. Diabetes 2007; 56(1): 3-7.
       [http://dx.doi.org/10.2337/db06-1165] [PMID: 17110468]

[75]   Atouf F, Park CH, Pechhold K, Ta M, Choi Y, Lumelsky NL. No evidence for mouse pancreatic β-cell epithelial-mesenchymal transition *in vitro.* Diabetes 2007; 56(3): 699-702.
       [http://dx.doi.org/10.2337/db06-1446] [PMID: 17327438]

[76]   Jamal A-M, Lipsett M, Sladek R, Laganière S, Hanley S, Rosenberg L. Morphogenetic plasticity of adult human pancreatic islets of Langerhans. Cell Death Differ 2005; 12(7): 702-12.
       [http://dx.doi.org/10.1038/sj.cdd.4401617] [PMID: 15818398]

[77]   Morton RA, Geras-Raaka E, Wilson LM, Raaka BM, Gershengorn MC. Endocrine precursor cells from mouse islets are not generated by epithelial-to-mesenchymal transition of mature β cells. Mol Cell Endocrinol 2007; 270(1-2): 87-93.
       [http://dx.doi.org/10.1016/j.mce.2007.02.005] [PMID: 17363142]

[78]   Ouziel-Yahalom L, Zalzman M, Anker-Kitai L, *et al.* Expansion and redifferentiation of adult human pancreatic islet cells. Biochem Biophys Res Commun 2006; 341(2): 291-8.

[http://dx.doi.org/10.1016/j.bbrc.2005.12.187] [PMID: 16446152]

[79]   Lechner A, Nolan AL, Blacken RA, Habener JF. Redifferentiation of insulin-secreting cells after *in vitro* expansion of adult human pancreatic islet tissue. Biochem Biophys Res Commun 2005; 327(2): 581-8.
[http://dx.doi.org/10.1016/j.bbrc.2004.12.043] [PMID: 15629153]

[80]   Russ HA, Bar Y, Ravassard P, Efrat S. *In vitro* proliferation of cells derived from adult human β-cells revealed by cell-lineage tracing. Diabetes 2008; 57(6): 1575-83.
[http://dx.doi.org/10.2337/db07-1283] [PMID: 18316362]

[81]   Weinberg N, Ouziel-Yahalom L, Knoller S, Efrat S, Dor Y. Lineage tracing evidence for *in vitro* dedifferentiation but rare proliferation of mouse pancreatic β-cells. Diabetes 2007; 56(5): 1299-304.
[http://dx.doi.org/10.2337/db06-1654] [PMID: 17303800]

[82]   Hanley SC, Pilotte A, Massie B, Rosenberg L. Cellular origins of adult human islet *in vitro* dedifferentiation. Lab Invest 2008; 88(7): 761-72.
[http://dx.doi.org/10.1038/labinvest.2008.41] [PMID: 18490899]

[83]   Russ HA, Sintov E, Anker-Kitai L, *et al.* Insulin-producing cells generated from dedifferentiated human pancreatic β cells expanded *in vitro*. PLoS One 2011; 6(9): e25566.
[http://dx.doi.org/10.1371/journal.pone.0025566] [PMID: 21984932]

[84]   Szabat M, Luciani DS, Piret JM, Johnson JD. Maturation of adult β-cells revealed using a Pdx1/insulin dual-reporter lentivirus. Endocrinology 2009; 150(4): 1627-35.
[http://dx.doi.org/10.1210/en.2008-1224] [PMID: 19095744]

[85]   Cox AR, Gottheil SK, Arany EJ, Hill DJ. The effects of low protein during gestation on mouse pancreatic development and β cell regeneration. Pediatr Res 2010; 68(1): 16-22.
[http://dx.doi.org/10.1203/PDR.0b013e3181e17c90] [PMID: 20386490]

[86]   Liu H, Guz Y, Kedees MH, Winkler J, Teitelman G. Precursor cells in mouse islets generate new β-cells *in vivo* during aging and after islet injury. Endocrinology 2010; 151(2): 520-8.
[http://dx.doi.org/10.1210/en.2009-0992] [PMID: 20056825]

[87]   Seaberg RM, Smukler SR, Kieffer TJ, *et al.* Clonal identification of multipotent precursors from adult mouse pancreas that generate neural and pancreatic lineages. Nat Biotechnol 2004; 22(9): 1115-24.
[http://dx.doi.org/10.1038/nbt1004] [PMID: 15322557]

[88]   Smukler SR, Arntfield ME, Razavi R, *et al.* The adult mouse and human pancreas contain rare multipotent stem cells that express insulin. Cell Stem Cell 2011; 8(3): 281-93.
[http://dx.doi.org/10.1016/j.stem.2011.01.015] [PMID: 21362568]

[89]   Razavi R, Najafabadi HS, Abdullah S, Smukler S, Arntfield M, van der Kooy D. Diabetes enhances the proliferation of adult pancreatic multipotent progenitor cells and biases their differentiation to more β-cell production. Diabetes 2015; 64(4): 1311-23.
[http://dx.doi.org/10.2337/db14-0070] [PMID: 25392245]

[90]   Beamish CA, Strutt BJ, Arany EJ, Hill DJ. Insulin-positive, Glut2-low cells present within mouse pancreas exhibit lineage plasticity and are enriched within extra-islet endocrine cell clusters. Islets 2016; 8(3): 65-82.
[http://dx.doi.org/10.1080/19382014.2016.1162367] [PMID: 27010375]

[91] Blaine SA, Ray KC, Anunobi R, Gannon MA, Washington MK, Means AL. Adult pancreatic acinar cells give rise to ducts but not endocrine cells in response to growth factor signaling. Development 2010; 137(14): 2289-96.
[http://dx.doi.org/10.1242/dev.048421] [PMID: 20534672]

[92] Herrera PL. Adult insulin- and glucagon-producing cells differentiate from two independent cell lineages. Development 2000; 127(11): 2317-22.
[PMID: 10804174]

[93] Thorel F, Népote V, Avril I, *et al.* Conversion of adult pancreatic α-cells to β-cells after extreme β-cell loss. Nature 2010; 464(7292): 1149-54.
[http://dx.doi.org/10.1038/nature08894] [PMID: 20364121]

[94] Yang Y-P, Thorel F, Boyer DF, Herrera PL, Wright CV. Context-specific α- to-β-cell reprogramming by forced Pdx1 expression. Genes Dev 2011; 25(16): 1680-5.
[http://dx.doi.org/10.1101/gad.16875711] [PMID: 21852533]

[95] Chung C-H, Hao E, Piran R, Keinan E, Levine F. Pancreatic β-cell neogenesis by direct conversion from mature α-cells. Stem Cells 2010; 28(9): 1630-8.
[http://dx.doi.org/10.1002/stem.482] [PMID: 20653050]

[96] Piran R, Lee S-H, Li C-R, Charbono A, Bradley LM, Levine F. Pharmacological induction of pancreatic islet cell transdifferentiation: relevance to type I diabetes. Cell Death Dis 2014; 5: e1357.
[http://dx.doi.org/10.1038/cddis.2014.311] [PMID: 25077543]

[97] White MG, Marshall HL, Rigby R, *et al.* Expression of mesenchymal and α-cell phenotypic markers in islet β-cells in recently diagnosed diabetes. Diabetes Care 2013; 36(11): 3818-20.
[http://dx.doi.org/10.2337/dc13-0705] [PMID: 24062329]

[98] Talchai C, Xuan S, Lin HV, Sussel L, Accili D. Pancreatic β cell dedifferentiation as a mechanism of diabetic β cell failure. Cell 2012; 150(6): 1223-34.
[http://dx.doi.org/10.1016/j.cell.2012.07.029] [PMID: 22980982]

[99] El-Gohary Y, Tulachan S, Wiersch J, *et al.* A smad signaling network regulates islet cell proliferation. Diabetes 2014; 63(1): 224-36.
[http://dx.doi.org/10.2337/db13-0432] [PMID: 24089514]

[100] Blum B, Hrvatin SS, Schuetz C, Bonal C, Rezania A, Melton DA. Functional β-cell maturation is marked by an increased glucose threshold and by expression of urocortin 3. Nat Biotechnol 2012; 30(3): 261-4.
[http://dx.doi.org/10.1038/nbt.2141] [PMID: 22371083]

[101] van der Meulen T, Xie R, Kelly OG, Vale WW, Sander M, Huising MO. Urocortin 3 marks mature human primary and embryonic stem cell-derived pancreatic α and β cells. PLoS One 2012; 7(12): e52181.
[http://dx.doi.org/10.1371/journal.pone.0052181] [PMID: 23251699]

[102] Blum B, Roose AN, Barrandon O, *et al.* Reversal of β cell de-differentiation by a small molecule inhibitor of the TGFβ pathway. eLife 2014; 3: e02809.
[http://dx.doi.org/10.7554/eLife.02809] [PMID: 25233132]

[103] Sabek OM, Ken Nishimoto S, Fraga D, Tejpal N, Ricordi C, Gaber AO. Osteocalcin effect on human β

cells mass and function. Endocrinology 2015; 156(EN): 2015-1143.
[http://dx.doi.org/10.1210/EN.2015-1143]

[104] Brown ML, Andrzejewski D, Burnside A, Schneyer AL. Activin enhances α- to β-cell transdifferentiation as a source for β-cells in male FSTL3 knockout mice. Endocrinology 2016; 157(3): 1043-54.
[http://dx.doi.org/10.1210/en.2015-1793] [PMID: 26727106]

CHAPTER 9

Induction of β-Cell Regeneration by Human Postnatal Stem Cells

Tyler T. Cooper, Ruth M. Elgamal and **David A. Hess**[*]

Molecular Medicine Research Group, Krembil Centre for Stem Cell Biology, Robarts Research Institute; Department of Physiology & Pharmacology, Schulich School of Medicine & Dentistry, Western University, London, Ontario, Canada

Abstract: The International Diabetes Federation estimates 382 million people are currently living with diabetes mellitus worldwide; and with increasing rates of obesity in an aging population this number is predicted to increase to 592 million by 2035. The inability to ameliorate the causes of diabetes has motivated researchers to develop novel approaches aimed at providing curative therapies to replace current symptomatic management using exogenous insulin. Accordingly, postnatal or adult stem cell transplantation has recently emerged as a promising therapeutic strategy following reports detailing the stimulation of islet regeneration in preclinical and early clinical studies. Postnatal bone marrow (BM) and umbilical cord blood (UCB) sources contain progenitor cells of hematopoietic, endothelial, and mesenchymal lineages; and each have demonstrated islet regenerative functions in animal models of diabetes. In the context of this chapter, we summarize accumulating evidence from preclinical and clinical studies describing transplantation of these specific postnatal lineages to stimulate the regeneration of endogenous insulin secreting β-cells, and how these stem cells may be used to provide paracrine support alongside the transplantation of allogeneic islets.

Keywords: Allogeneic transplantation, Autologous transplantation, β-cells, Bone marrow, Diabetes mellitus, Endothelial progenitor cells, Hematopoietic progenitor cells, Hypoxia, Insulin, Islet angiogenesis, Islet neogenesis,

[*] **Corresponding author David A. Hess:** Molecular Medicine Research Group, Krembil Centre for Stem Cell Biology, Robarts Research Institute, Department of Physiology & Pharmacology, Western University, 1151 Richmond St, London, Ontario, Canada, N6A 5B7; Tel/Fax: 519.931.5777x.24118; E-mail: dhess@robarts.ca

David J. Hill (Ed.)

Islets of langerhans, Islet regeneration, Multipotent stromal cells, Pancreas, Paracrine signals, Progenitor cells, Stem cells, Transplantation, Umbilical cord blood.

INTRODUCTION

Diabetes mellitus is generally characterized by the body's inability to maintain controlled glycemic levels due to pancreatic β-cell death or dysfunction [1]. Type 1 diabetes (T1D), also referred to as "juvenile onset diabetes" or "insulin-dependent diabetes", arises as a result of the autoimmune destruction of insulin-secreting β-cells in the islets of Langerhans. T1D is mediated by an inflammatory infiltrate composed of $CD4^+$ and $CD8^+$ T-lymphocytes, B-lymphocytes, macrophages, and NK-cells [2, 3]. β-cell-specific autoimmune depletion leaves T1D patients with an inadequate supply of endogenous insulin, requiring exogenous insulin to control hyperglycemia. In contrast, T2D, often referred to as "late-onset diabetes" or "non-insulin-dependent diabetes", occurs as a result of insulin insensitivity or resistance in peripheral tissues (skeletal muscle, liver, and adipose), and is associated with a combination of risk factors including obesity, sedentary lifestyle, environmental stimuli, and/or genetics [4]. In the early stages of T2D, prolonged hyperglycemia stimulates β-cells to over secrete insulin and results in β-cell exhaustion and apoptosis [5], culminating in reduced β-cell mass as T2D develops [6, 7]. Regardless of these pathological differences, T1D and T2D, ultimately result in the loss of β-cell mass over time; leading to inadequate insulin secretion within islets and the requirement for exogenous insulin therapy. Due to the insulin deficiency created by diabetes, restoration of physiological insulin secretion is essential to treating both T1D and T2D and its subsequent complications.

Dealing with the Complications and Consequences of Diabetes Mellitus: An Emerging Global Crisis?

In 2013, the International Diabetes Federation (IDF) estimated 382 million people are currently living with diabetes mellitus worldwide; and this number is predicted to increase to 592 million by 2035 [4, 8]. Specifically, >90% of diagnosed individuals have T2D, while <10% have T1D. Due to increasing rates of obesity and aging population demographics worldwide, diabetes mellitus has reached pandemic proportions. Since the discovery of insulin by Banting and Best in the early 1920s [9], and despite the recent development of improved insulin administration (automated insulin pumps) [10], both T1D and T2D patients ultimately develop serious comorbidities and secondary complications associated with inadequate control of glycemia. The most severe complications include increased rates of cardiovascular diseases (heart attack, coronary and peripheral artery disease, stroke), vision loss/blindness, kidney failure, nerve damage, problems with pregnancy, and depression. According to the Public Health Agency of Canada, individuals with diabetes are 3 times more likely to develop cardiovascular disease, 12 times more likely to be hospitalized with end-stage renal failure, and 20 times more likely to require non-traumatic lower limb amputation [11]. Although it is difficult to assess the economic burden of diabetes worldwide, the American Diabetes Association estimated that the total cost of diabetes in 2012 exceeded $245 billion in the USA ($176 billion in direct medical costs and $69 billion in reduced productivity) [12]. These staggering numbers have prompted researchers to aggressively investigate improved and potentially curative therapies to combat both T1D and T2D.

The Advent of Cellular Therapies for Diabetes: The Edmonton Protocol

Recently, cell-based therapies have emerged as a frontrunner to provide curative therapies for diabetes using islet replacement to restore functional β-cell mass [13]. In 2000, the pioneering efforts of the Edmonton protocol provided 'proof-o--concept' that portal vein transplantation of cadaveric human islets, combined with modern immunosuppressive therapy, could lead to insulin independence in patients with severe T1D [13, 14]. Although results at 1 year were initially promising with 7 of 7 patients achieving insulin-independence after the transplantation of islets from at least 2 donor pancreata, islet rejection and continued autoimmune assault resulted in the return to insulin therapy in the majority of patients [13]. In addition, islet survival and function was compromised by isolation procedures and engraftment in the hepatic site. Widespread application of this approach remains limited due to an extreme shortage of donor pancreas tissue available for transplantation. Finally, due to the requirement for life-long immunosuppression and increased risk of infections, islet transplantation is only indicated for brittle T1D patients [15].

With these limitations in mind, two broad approaches are currently under intense pre-clinical investigation to provide curative therapies for patients with T1D. First is the generation of an unlimited supply of exogenous β-cells/islets derived from pluripotent (hES of iPS cells) sources for replacement therapies. Although recent advancements have been achieved in the derivation of β-like cells from hES cells [16 - 18], clinical implementation establishing safety and efficacy of this approach are currently under considerable scrutiny [19]. Second is the induction of *endogenous* islet regeneration in diabetic patients, using the transfer of islet regenerative cells or stimuli in combination with immunosuppressive therapies. Because the derivation and transplantation of β-cells from pluripotent sources has been covered in a previous chapter in this book, this chapter will focus on pre-clinical and clinical strategies to administer postnatal stem cells to "tip the balance" in favour of islet regeneration *versus* destruction during diabetes.

Can Islet Regeneration Occur in the Face of Autoimmunity: The Medalist Study?

In 2010, Keenan *et al.* studied ß-cell function in patients that have lived with T1D for >50 years with minimal complications; also known as the Joslin Medalist cohort. Most of these long-term survivors demonstrated C-peptide production indicating that residual insulin was continually synthesized during diabetes. Furthermore, analysis of 9 Medalist pancreata post-mortem revealed the presence of proliferating insulin$^+$ β-cells, suggesting islet regeneration could occur in the face of ongoing autoimmunity [20]. Moreover, the increase in β-cell mass in response to pregnancy [21, 22] or obesity during the early stages of T2D [23] suggests the human endocrine pancreas has the capacity to regenerate. These studies provide hope for the future development of pharmaceutical, protein, and/or cellular therapies that could be used to stimulate proliferation and insulin production within residual β-cells to treat or reverse established diabetes. Thus, novel therapeutic strategies to optimize islet regeneration within the diabetic pancreas are under intense investigation.

Bone Marrow Stem Cells Initiate Islet Regeneration: Identifying the Mechanisms?

In one of the first publications on endogenous islet regeneration, we demonstrated that transplantation of murine BM-derived, c-kit$^+$ stem cells reduced hyperglycemia in mice with streptozotocin (STZ)-induced β-cell deletion [24]. Importantly, transplanted eGFP+ donor cells did not acquire insulin expression but engrafted the pancreas in ductal regions and surrounding islets; and stimulated the proliferation of recipient islets *via* undetermined paracrine activities. In the same year, Ianus *et al.* reported results that BM-transplantation using male donors where eGFP was expressed under the control of the rat insulin promoter, produced rare Y-chromosome/eGFP$^+$ cells within islets without evidence of cell fusion [25]. The authors concluded that the mouse BM contained cells that could differentiate into glucose-responsive β-cells, adding fuel to the fire surrounding the application of stem cell "transdifferentiation" as a clinically relevant concept for cell-based therapies. In light of these conflicting results, follow-up reports using similar transplantation systems quickly surfaced citing little to no evidence for the transdifferentiation of BM cells into pancreatic β-cells *in vivo* [26, 27]. Nonetheless, BM-derived stem and progenitor cells were capable of improving islet function after transplantation and the search was on to identify human cells that possessed islet regenerative capacity after transplantation.

Subsequently, our group and others have shown that human hematopoietic progenitor cells (HPC), multipotent-stromal or mesenchymal stem cells (MSC), and endothelial progenitor cell (EPC), could enhance islet regeneration after pancreas damage *via* various mechanisms [28]. In addition, islet regeneration and immune protection could be achieved by co-infusion of murine hematopoietic cells with MSC in autoimmune NOD mice [29, 30]. Although this concept termed 'stem cell-stimulated islet regeneration' has emerged as a central process for pancreas repair, controversy has arisen surrounding the mechanisms by which post-natal stem cells mediate islet regeneration, either by stimulating β-cell proliferation [31, 32], or by initiating islet neogenesis from ductal or islet-derived precursors [33 - 35]. Although both processes may be harmonized during functional islet repair, the mechanisms by which distinct progenitor lineages mediate islet recovery are poorly understood, and remain the key to translating the ability of post-natal stem cells to induce islet regeneration into a clinically relevant treatment for diabetes.

In the context of this chapter, we will focus on pre-clinical and clinical studies transplanting progenitor cells from the hematopoietic, mesenchymal, and endothelial cell lineages (Fig. **1**), in order to dissect the emerging regenerative and immunomodulatory mechanisms involved.

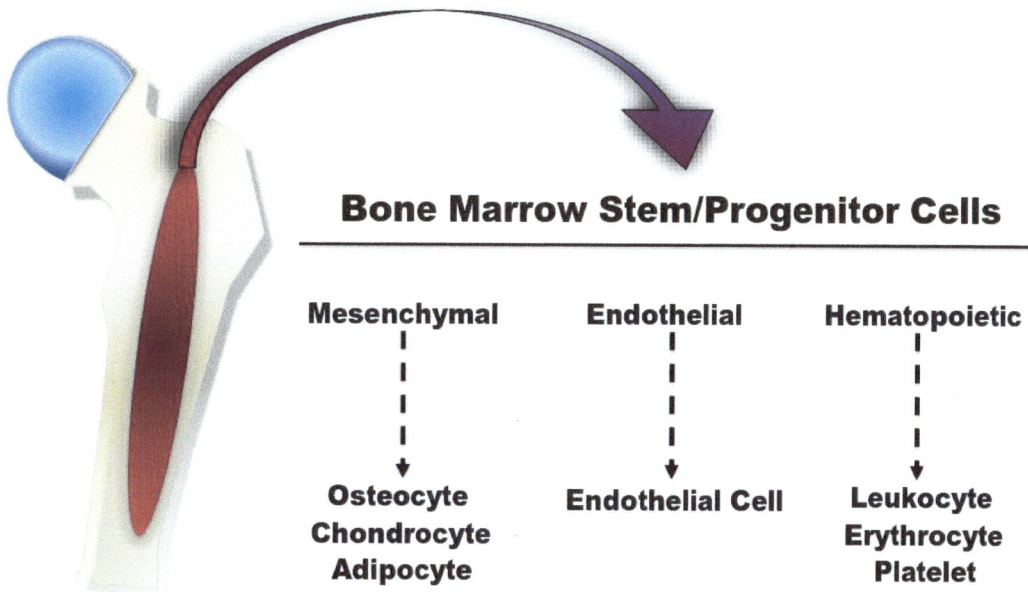

Bone Marrow Stem/Progenitor Cells

Mesenchymal	Endothelial	Hematopoietic
↓	↓	↓
Osteocyte Chondrocyte Adipocyte	Endothelial Cell	Leukocyte Erythrocyte Platelet

Fig. (1). Human bone marrow contains several lineages of stem/progenitor cells with islet-regenerative functions. Putative lineages with islet-regenerative functions include (1) pro-regenerative hematopoietic stem and progenitor cells (HSPC) that stimulate β-cell proliferation (2) multipotent/mesenchymal stromal cells (MSC) that induce islet formation associated with the ductal epithelium, and (3) pro-angiogenic endothelial progenitor cells (EPC) involved in islet capillary formation and re-vascularization.

HEMATOPOIETIC STEM AND PROGENITOR CELLS

BM-derived hematopoietic stem and progenitor cells (HSPC) represent the most highly studied stem-cell population since their initial discovery by Till and McCulloch in the 1960s [36]. HSPCs inherently possess the capacity to differentiate into mature erythroid, myeloid, and lymphoid cell types, and to establish a hierarchy within the BM in both murine [37] and human systems [38].

At the apex of the hierarchy, the most primitive HSC are thought to self-renew in association with the endosteal surfaces of the BM. Generally, more differentiated HPC reside in the perivascular niche [39]; where more committed cell progeny have access to the peripheral circulation in response to chemokines and growth factor stimuli, allowing them to home to sites of injury [40].

Using transplantation into highly immunodeficient (NOD/SCID) recipients, the capacity for HSPC to reconstitute human hematopoiesis within the BM has been clearly established. The expression of the glycoprotein saliomucin, or CD34, has been used to quantify a cell's capability to mediate hematopoietic reconstitution in mice, primates, and humans following myleoablation [41]. As HSPC mature, the expression of CD34 is diminished, giving rise to lineage-restricted CD34$^-$ cells with reduced repopulating function. CD34$^+$ HSPC represent only ~1.5% of human BM mononuclear cells, but are highly amenable to *ex vivo* expansion under defined, clinically applicable conditions to generate more cells for clinical therapies. Notably, CD34$^+$ HSPC do not reside at the apex of hematopoiesis as a rare highly quiescent population of CD34$^-$/CD38$^-$/CD93$^+$ cells have recently demonstrated delayed, yet extensive long-term and repopulating capacity and the ability to generate CD34$^+$ HSPC after serial transplantation [42]. Although CD34$^-$expression is largely accepted as the standard for HSPC enumeration during current BM-transplantation therapies, co-expression of several other cell surface markers (CD133, CD117, CD45RA, CD90, or CD49f) in combination with CD34 expression may purify more primitive populations with greater regenerative potential in alternate indications [43].

Why Use HPC for Islet Regeneration?

Clinical use of HSPC in transplantation procedures is not limited to the reconstitution of the blood system during hematopoietic disorders. Recently, HPC have also demonstrated the ability to promote blood vessel formation without directly incorporating into new vessels in murine models of critical limb ischemia [44]. HPC exert their proangiogenic effects by recruiting host endothelial cells and platelets to sites of injury, activating a pro-angiogenic switch in a paracrine manner [40]. As such, the pro-angiogenic functions of HPC can be used to support revascularization of regenerating islets or co-transplanted islet allografts.

Hasawega *et al.* demonstrated that after BM-transplantation in irritated mice, the mobilization of donor cells from the recipient BM can promote islet function after β-cell ablation using STZ [45]. Importantly, mobilized (eGFP+) donor cells homed to regenerating islets and predominantly expressed the pan-leukocyte marker CD45. Moreover, purified (CD34+) HSPC demonstrate the ability to significantly expand under clinically applicable conditions while limiting their differentiation through fed batch [46 - 48] and/or pathway manipulation with growth factors [49, 50]; in return increasing the number of cells for clinical applications. Nonetheless, HSPC can be isolated from alternative sources such as mobilized-peripheral blood (MPB) or UCB allowing the development of autologous and allogeneic strategies towards therapeutic application in diabetes [51].

Isolation of Human Hematopoietic Progenitor Cells Using High ALDH-Activity

In order to purify stem cells from human BM and UCB, a fluorescent substrate (Aldefluor™) of ALDH, a conserved detoxification enzyme highly expressed in long-lived progenitor cells, can be used [52]. Interestingly, high ALDH-activity activates intracellular retinoic acid (RA) signalling, a pathway that paradoxically induces differentiation in HPC [49, 50, 53]. Thus, as HPC mature, ALDH-activity is reduced. Importantly, the Aldefluor™ substrate is non-toxic and efficiently cleared from cells after purification. Thus, fluorescence activated cell sorting (FACS) for high ALDH-activity represents a clinically applicable approach to purify cells for expansion and transplantation. ALDH^{hi} cells represent a rare fraction of human BM (<0.8%) [54] or UCB (<0.5%) [55] and are primarily hematopoietic (>90% CD45+) in origin [56 - 58]. ALDH^{hi} cells from both sources co-express stem cell-associated surface markers (CD34, CD133), efficiently form hematopoietic colonies *in vitro*, and demonstrate hematopoietic repopulating function after transplantation into sublethally-irradiated, immunodeficient NOD/SCID mice [57, 58]. In addition, ALDH^{hi} cells promote revascularization in humanized murine models of hind limb ischemia. Notably, ALDH^{hi} cells are highly proliferative in serum free, xeno-free liquid cultures under hematopoietic-restricted conditions optimized to prevent HPC differentiation. Thus, ALDH^{hi} cells generate a large number of hematopoietic progenitor cells (HPC) for the rational development of islet regenerative therapies.

Preclinical Xenotransplantation Models to Investigate Islet Regeneration

We have used multiple low-dose STZ-injections in NOD/SCID mice to characterize islet regeneration after transplantation human ALDH[hi] cells or culture expanded HPC [59 - 61]. Importantly, this assay represents a humanized model to study the induction of islet regeneration after chemically induced β-cell ablation, as the NOD/SCID model demonstrates reduced innate immunity and NK-cell activity (due to the NOD mutation), and lacks T and B-lymphocyte function (due to the SCID mutation). Although this model is described briefly in this chapter, for details on our methods to characterize mechanisms of stem cell-induced islet regeneration after systemic (intra-venous) and direct intra-pancreatic transplantation *in vivo*, please refer to our published review [62].

Human HPC with High ALDH-activity Promote Regenerating Islet Cell Proliferation and Vascularization

Transplantation of freshly-isolated BM ALDH[hi] cells consistently improved hyperglycemia *via* the recovery of β-cell mass [59]. Notably, transplanted ALDH[hi] cells increased islet size by augmenting cell proliferation and vascular density within regenerating islets, without a significant increase in total islet number. Thus, ALDH[hi] cells stimulated regeneration *via* the paracrine induction of an islet proliferative program *in situ*. Global genomic and proteomic analyses to identify specific cytokines secreted by ALDH[hi] cells that impact β-cell proliferation are currently underway. However, iv-injected ALDH[hi] cells showed limited and only transient recruitment to the pancreas [59, 60]. Therefore, we developed simple micro-injection methods for the delivery of cells directly into the pancreas (iPan-injection) [61]. Compared to iv-injection with ALDH[hi] cells from matched UCB samples, mice iPan-injected showed augmented recovery from hyperglycemia, improved glucose tolerance, and more robust induction of islet-cell proliferation. Indeed, at 4 and 7 days post-transplantation, ALDH[hi] cells surrounded regenerating islets and stimulated the proliferation of β-cells or putative β-cell precursors as insulin expression gradually recovered. These studies demonstrated the utility of UCB-derived ALDH[hi] hematopoietic cells as a novel starting population for the development of islet regenerative therapies. Due to documented progenitor cell dysfunction in patients with diabetes and vascular co-morbidities

[62 - 64], the use of autologous BM cells for islet regeneration is predicted to show reduced regenerative function. Fortunately, recent government initiatives to bank and HLA-phenotype UCB samples for clinical applications has established UCB as a readily available source of allogeneic cells, early in ontogeny and untouched by disease-related pathologies. Also, after iPan-transplantation of fresh UCB ALDH[hi] cells, permanent engraftment was not required to effectively stimulate an islet proliferative program. Thus, future allogeneic regimens transplanting expanded UCB HPC subsets during T1D may only require short-term immunosuppression to stimulate a long-lasting islet regenerative response.

Islet Regenerative Functions by Expanded HPC Subsets with High ALDH-Activity

Clinical use of BM or UCB ALDH[hi] cells to treat diabetes will require the expansion of purified HPC without compromising islet regenerative functions. We have developed clinically applicable culture protocols that promote HPC expansion while minimizing differentiation and have designed purification strategies to re-select expanded HPC that retain the islet regenerative (ALDH[hi]) phenotype. Using the principle that ALDH-activity is decreased as progenitor cells differentiate [49, 52, 54, 55, 57], we re-purified expanded HPC based on low *versus* high ALDH-activity [65]. After 6 days expansion, a 20-fold increase in total UCB cells was achieved; however, injection of the bulk expanded HPC population did not reverse hyperglycemia after transplantation into hyperglycemic mice. Interestingly, re-purification of ALDH[hi] HPC stimulated β-cell proliferation and reduced hyperglycemia similar to freshly isolated ALDH[hi] cells. In contrast, transplantation of re-purified ALDH[lo] did not demonstrate islet regenerative capacity.

Although promising, expansion of ALDH[hi] HPC becomes a paradoxical challenge, as ALDH is the rate-limiting enzyme in the production of the potent morphogen retinoic acid (RA) [66]. Upon its production, RA will translocate and activate the nuclear retinoic acid receptor complex (RAR/RXR) and binds to promoter regions upstream of numerous targets involved in hematopoietic cell differentiation [67]. As a result, ALDH[hi] UCB cells are intrinsically driven to differentiate, thus diminishing the frequency of pro-regenerative progenitor cells during *ex vivo* expansion. Thus, future studies will focus on pharmacological RA-pathwa--inhibition to conserve regenerative functions while preventing unwanted differentiation. Additionally, identification of surface markers and secreted effectors that mediate islet regenerative functions is highly warranted.

Can HPC Transplantation Abrogate Autoimmunity and Permit Endogenous Islet Regeneration?

Implementation of cellular therapies designed to stimulate the recovery of endogenous β-cells will also require the abrogation of autoimmunity underlying the initial pathogenesis of TID. In essence, regenerated islets must be able to survive and function in the midst of ongoing autoimmunity. In the early 1990s, Sibley *et al.* reported the emergence of T1D in a female patient several years after receiving a hematopoietic stem cell transplantation (HSCT) from her HLA-identical brother with T1D [68]. This report identified lymphopoietic dysregulation within adoptively transferred donor cells as a pivotal component in the pathogenesis of T1D. However, reversal of this concept may also apply, where reconstitution of hematopoiesis from healthy marrow may abolish the immunological intolerance inherent in T1D recipients. Several studies have demonstrated that allogeneic HSCT could delay the onset of diabetes in NOD mice but provided minimal benefits after diabetes was established [69]. On the other hand, Wen *et al.* demonstrated that allogeneic BM-HSCT performed within 10 days after the onset of overt diabetes could reverse hyperglycemia in NOD mice [70]. The authors suggested the induction of prolonged remission was largely mediated by the expansion of $CD4^+CD25^+FoxP3^+$ regulatory T-lymphocytes. Although preliminary, these studies provide fundamental evidence to suggest the re-establishment of hematopoiesis from HSPC following myeloablative conditioning may replace autoreactive T-lymphocytes responsible for the destruction of pancreatic islets.

In the early 2000s, Voltarelli and colleagues demonstrated insulin independence for up to 35 months in 12 of 20 newly diagnosed T1D patients who underwent non-myeloablative autologous BM transplantation collected from G-CSF mobilized peripheral blood [71]. In a follow up study, Couri *et al.* documented 12 of 20 patients remained insulin independent for >14 months with increased circulating C-peptide and reduced serum-hemoglobin A1C (HbA$_{1C}$) [72]. Finally, Haller *et al.* conducted a phase I/II study where newly T1D patients that (<6 weeks post-diagnosis) underwent autologous UCB-HSCT, without any myeloablative pre-conditioning, exhibited an increase in CD4$^+$CD25$^+$FoxP3$^+$ regulatory T-lymphocytes and delayed the progression of insulin dependence [73]. Although, these studies provided evidence for the safety and efficacy of HSCT for diabetes therapy, complete and sustained reversal of hyperglycemia has not been fully demonstrated. Nonetheless, allogeneic HSCT in combination with HLA-matched islet or pancreas transplantation, may represent a future strategy to achieve sustained reversal of diabetic symptoms.

MULTIPOTENT MESENCHYMAL STROMAL CELLS (MSC)

In the 1970s, Friedenstein and colleagues first identified clonogenic, plastic adherent cells from the BM that they termed colony forming units of fibroblasts (CFU-F) [74]. Friedenstein further documented the capacity of CFU-F to differentiate into canonical cell-types of the mesodermal lineage including osteocytes, chondrocytes and adipocytes; however, did not possess the ability to generate other mesoderm-derived cell-types such as blood. As a result, many researchers have subsequently characterized this multipotent-stromal cell population to understand their biological and molecular functions useful for the development of develop novel cellular therapies.

These cells were first termed marrow stromal stem cells, or most accurately multipotent-stromal cells (MSC), based on their differentiative potentials. In 1991, the term mesenchymal stem cells (also MSC) was coined by Arnold Caplan during studies which highlighted their putative developmental origin and extensive therapeutic potential [75]. Promiscuous use of these related nomenclatures has sparked confusion and debate in the field because the stem cell nature and self-renewal potential of MSC remains highly controversial [76, 77]. In 2006, the International Society of Cellular Therapies (ISCT) provided standardized criteria for MSC based on 3 essential criteria [78]. (1) Prolonged culture adherent to plastic generates a non-hematopoietic (CD45-negative) population with a fibroblast-like (spindle-shaped) morphology. (2) Under serum containing culture conditions, these cells are highly express stromal cell surface markers such as CD73, CD90, and CD105, and are devoid of hematopoietic markers including CD45, CD34, CD14, CD19, and HLA-DR. As such, the cell surface expression profile established by the ISCT is a widely-acknowledged standard for defining the purity of MSC during ex *vivo* culture [78]. (3) The cell progeny generated must efficiently differentiate into mature osteocytes, chondrocytes and adipocytes under the appropriate culture conditions. Unfortunately, identification of uniquely-expressed cell surface markers has remained elusive. Nonetheless, MSC can be isolated from multiple human tissues including the bone marrow, kidney liver, adipose, placental, and muscle tissues. Recently, Crisan *et al.* has established that MSC originate from pericytes and adventitial cells in developing and adult human organs, wrapping endothelial cells, capillaries and arteries [79]. Importantly, these MSC expressed CD146 and PDGFRβ *in situ*, without the expression of hematopoietic (CD34, CD45) or endothelial (CD31, vWF) markers; allowing for the first time a rational selection strategy to reduce the heterogeneity of this population prior to culture expansion. Importantly, these cells fulfill all the defining criteria established by the ISCT, and differentiate towards muscle, bone, adipose and cartilage lineages, both *in vitro* and after transplantation *in vivo*.

Within the BM, MSC possess an innate ability to sense/ respond to the microenvironment and enable paracrine control over proliferation, differentiation, and/or migration of resident hematopoietic and endothelial cells [80]. The regenerative capabilities of MSC in murine disease models are attributed to cross-talk between MSC and endogenous cell populations by primarily unidentified signalling pathways [81]. Considering the innate differentiation capacity of MSC *in vitro* and *in vivo* [82], numerous labs have investigated the capacity for MSC to transdifferentiate into β-cells. Once again, their ability to produce functional, insulin-producing β-like cells has only been demonstrated *in vitro* under manipulated culture conditions or after transfection [83]. Nonetheless, MSC remain an attractive cell type for the development of regenerative therapies for diabetes due to the following characteristics. (1) MSC are readily availability in a variety of tissues; (2) MSC are capable of reaching the large numbers required for cellular therapies; (3) MSC secrete a vast array of pro-angiogenic and regenerative cytokines in response to tissue damage; (4) MSC demonstrate immunomodulatory properties and evade immuno-surveillance in the allogeneic setting. MSC may be able to provide local protection from autoimmunity after transplantation in multiple organs [82, 84]. Thus, MSC are uniquely positioned for the development of novel cell therapies to treat diabetes, including the paracrine support of endogenous islet regeneration, and potential islet-protective dampening of autoimmunity.

Can MSC Create a Regenerative Niche for New Islet Formation?

In 2006, Lee *et al.* were the first to establish that human BM-derived MSC transplantation could combat the symptoms of diabetes mellitus and its secondary complications [85]. In STZ-treated NOD/SCID mice, MSCs were transplanted *via* intracardiac infusion, to avoid lodging and engraftment in the lung and liver. Increased murine insulin coincided with a reduction in blood glucose in mice receiving MSC infusions, and suggested the stimulation of endogenous islet regeneration. Histological analyses confirmed increased insulin expression within small islet-like structures associated with the terminal ductal-epithelium. Our group has independently confirmed and extended these findings using BM-MSC derived initially from the non-hematopoietic fraction of ALDH^{hi} cells. However, distinct MSC samples showed variable capacity to reduce hyperglycemia with ≈40% of MSC samples showing measurable islet regenerative capacity. However, transplantation of the regenerative MSC samples also increased the number of small islets associated with ducts without stimulating β-cell proliferation [59, 60]. Furthermore, developing islets contained an increased frequency of intra-islet ductal epithelial cells, and iPan-injection of MSC resulted in the emergence of single insulin-positive cells surrounding ducts at 7 days post-transplantation. These data suggested that transplanted MSC stimulated a putative islet neogenic mechanism. However, histological sections were devoid of Ngn3 expression, a marker of endogenous endocrine progenitors that contribute to islet formation during development or after partial pancreatectomy. Diabetic nephropathy, a common cause of end-stage renal failure in T1D patients; was also reduced in this model *via* decreased mesangial thickening and pro-inflammatory macrophage infiltration. These studies established the possibility that MSC may provide therapeutic benefits for secondary complications in addition to supporting endogenous islet regeneration during diabetes [86].

Several research labs have demonstrated that MSC possess the ability to home to injured tissues/organs in a chemokine-dependent fashion. Sordi *et al.* demonstrated this concept *in vitro* using the supernatant of cultured STZ-treated islets. MSC expressing CXCR4, CXCR6, CCR1, and CCR7 were recruited to the injury stimulus by sensing; CX3CL1 and CXCL12 [87]. Similarly, Cheung *et al.* transplanted MSC through the tail-vein of STZ-induced diabetic NOD/SCID mice after iPan injection of stromal-cell derived factor 1 (SDF-1 or CXCL12) [88]. Interestingly, SDF-1 increased the presence of MSC within the pancreas, specifically in areas with high blood vessel density. Overall, the ability of MSC to sense, migrate, and respond to damage increases their therapeutic potential for the treatment of diabetes and its secondary complications.

MSC are highly secretory and release a broad range of growth factors to impact local repair processes. Interestingly, Lee *et al.* extended their initial contributions to describe MSC-derived anti-inflammatory proteins released into the blood stream could stimulate repair of damaged myocardium, despite MSC engraftment in the lungs [89]. This concept could be extended as a mechanism to support the expansion of β-cell mass by the secretion of regenerative factors (VEGF, TGF-β, EGF, *etc.*) or hormones without direct interaction in the pancreas [86, 90]. In support of this concept, Gao *et al.* demonstrated the injection of MSC-generated conditioned media alone could induce a significant increase in the number of endogenous islets and β-cells [91]. Moreover, the islet recovery was reversed by pharmacological inhibition of Akt-signalling but not Mek/Erk pathways *in vivo*. In a study by Si *et al.*, increased β-cell function and increased insulin sensitivity was reported when MSC were transplanted into T2D mouse models (STZ + high fat diet). Although benefits were only sustained for 4 weeks, subsequent transplantation again produced transient improvements, highlighting the potential for repeated MSC-infusion for the treatment of T2D [92]. Further research is required to elucidate the molecular mechanisms underlying the communication between MSC and regenerating β-cells for the development of cell-based therapies.

Although pre-clinical justification for the use of MSC transplantation to treat T1D is accumulating, the regenerative and proliferative capacity of MSC has been documented to vary based on the extent of expansion and the age or health of donors [93, 94]. Thus, a major challenge moving towards clinical implementation will be the identification of regenerative subsets within heterogeneous MSC populations, and how secreted effectors are impacted by donor variability and passage number [60, 95]. Our previous studies attempted to segregate the regenerative capabilities of BM-derived MSC subsets based on the activity of ALDH. Interestingly, individual BM-samples varied considerably in their capacity to regenerate endogenous islets, regardless of further purification into ALDH^{hi}*versus* ALDH^{lo} subsets. Furthermore, islet regenerative capacity was diminished with prolonged expansion. Although mRNA expression profiles revealed transcriptional differences between regenerative and non-regenerative subsets; a distinct molecular signature for regenerative MSC remains elusive and will require detailed secretome analyses [60].

Given the considerable potential surrounding MSC-transplantation for treating diabetes, several clinical trials have been conducted or are currently underway in this sector. In a randomized controlled trial, Hu *et al.* evaluated the long-term effects of transplanting newly-diagnosed T1D patients with allogeneic MSC derived from Wharton's Jelly. Importantly, the authors reported both HbA1c and C-peptide levels were significantly improved from baseline. Moreover, no acute or chronic side effects were reported, suggesting the use of MSC is safe for the treatment of T1D [96]. In a pilot-study, 10 patients with long-standing T2D implanted with allogeneic placenta-derived MSC showed reduced insulin requirements and increased C-peptide at 3 months [97]. Several trials involving the transplantation of autologous BM-derived MSC (NCT01157403 and NCT02057211), allogeneic BM-derived MSC (NCT00690066), or autologous adipose-derived MSC (NCT00703599) are currently underway. Considering MSC may provide benefits for secondary complications of diabetes, trials focused on utilizing MSC for diabetic-induced nephropathy (IRCT201111291414N28), critical limb ischemia (NCT00955669 and NCT01065337), and foot ulcers (NCT01216865) have also been launched. Continuing translation of emerging pre-clinical insight detailing islet-regenerative mechanisms stimulated by MSC will allow for better assessment of safety and effectiveness of new cellular therapies for patients with T1D and T2D.

MSC Immunomodulation and Recruitment of M2 Macrophages to Promote Islet Regeneration

In 2003, Li *et al.* showed BM-transplantation reduced endocrine destruction in a pancreatitis-induced diabetic mouse model using transgenic mice deficient in the transcription factors E2f1 and E2f2, transcriptional regulators of cell cycle proteins [98]. This led to speculation that a pro-regenerative cellular subset within unfractionated BM that harbors islet-protective capacity in a pro-inflammatory microenvironment. In a follow-up study, this group went on to demonstrate that M2 macrophages migrated to the damaged pancreas of E2f1/E2f2 knockout mice [99] and led to angiogenesis and proliferation within murine islets after transplantation of wild-type BM, whereas BM from macrophage-deficient colony stimulating factor-1 receptor (Csf1r$^{-/-}$) mice did not provide a measurable benefit. In an independent study utilizing a pancreatic ductal ligation model (PDL), Xiao *et al.* was able to demonstrate the islet-regenerative mechanisms initiated by M2 macrophage secretion of transforming growth factor-1 (TGFβ1) and epidermal growth factor (EGF), respectively [100]. To confirm these molecular changes were induced by M2 macrophages, the authors induced macrophage apoptosis by injecting PDL mice with clodronate, a myeloid-ablating liposome. Protein levels of SMAD7, CyclinD1/D2 within β-cells significantly reduced with proliferation, thus supporting that M2 macrophages can support β-cell regeneration. To extend the M2 macrophage paradigm, transplanted MSC may recruit pro-regenerative cell populations from the circulation through chemokine signals in addition to directly supporting islet regeneration. Cao *et al.* demonstrated in STZ-induced diabetic mice that transplanted MSC recruited and polarized macrophages to a pro-regenerative M2 phenotype in an SDF-1/CXCR4-dependent fashion. This cascade was reversed by administration of AMD3100, an agent that blocks the SDF-1/CXCR4 signalling axis [101]. Although subsequent research is needed to elucidate the underlying mechanisms involved in MSC-mediated recruitment and polarization of macrophages, successful development of curative therapies for T1D may entail combinations of regenerative MSC transplantation with secondary cell populations such as immunomodulatory M2 macrophages.

Can MSC Reverse Autoimmunity in T1D and Support Endogenous β-Cell Regeneration?

In addition to their pro-regenerative functions, the immunomodulatory properties of MSC may emerge as beneficial for the sustained reversal of diabetes. The immunological intolerance and subsequent destruction of β-cells in T1D is mediated by $CD4^+$ T-cells, $CD8^+$ T-cells, NK cells, and pro-inflammatory macrophages [3]. Previous studies have demonstrated that inhibition of lymphocyte function and proliferation can temporarily support insulin independence in patients with T1D [102]. Furthermore, Takashi *et al.* demonstrated the delivery IL-10 and a pro-insulin auto-antigen, in combination with CD3 blocking antibodies, reversed hyperglycemia in newly diabetic NOD mice [103]. These studies have provided the initial 'proof-of-concept' that abolishment of autoimmunity or induction of immune tolerance by MSC, in parallel with restoration of β-cell mass, harbors the potential to provide a cure for diabetes. MSC secrete anti-inflammatory/pro-regenerative cytokines (IL-10, TGF-B, IDO, and PGE2) which dampen autoimmune responses by repressing T-, B-, and NK-cell activation, maturation, and proliferation [104 - 107]. Collectively, the potential of MSC to support β-cell regeneration, in addition to generating a protective niche against autoimmune β-cell destruction, represents an exciting opportunity to explore. In 2009, Fiorina *et al.* demonstrated allogeneic transplantation of BALB/C-derived MSC could prevent the onset of diabetes in prediabetic NOD mice; however, this protection was not observed following the autologous NOD-MSC transplantation [104]. In a similar study, Madec *et al.* suggested that autoimmune protection in NOD mice was mediated by MSC activation of $Foxp3^+$ regulatory T-lymphocytes [108]. Boumaza *et al.* further documented an increase of IL-10/13 production by regulatory T- lymphocytes ($CD4^+/CD8^+/Foxp3^+$) in MSC-transplanted rats [109]. Based primarily on the immunomodulatory properties of MSC for diabetic therapies, the Prochymal trial (NCT00690066) was launched to determine if infusion of allogeneic MSC can halt autoimmunity and restore glycometabolic homeostasis in newly diagnosed diabetic patients. However, the preliminary results from this trial have not yet been reported. Finally, Zhao *et al.* developed a procedure coined the "Stem Cell Educator" therapy [102], where the peripheral blood of 12 patients with T1D was circulated through a closed-loop system exposed to adherent UCB-derived stem cells prior to reinfusion. Although the makeup of these 'educator cells' was not well characterized, patients reinfused with their 'educated' cells showed improved levels of HbA_{1c}, circulating C-peptide, and required less daily insulin for up to 6 months duration. The authors reported an increase in the number of $CD4^+CD25^+Foxp3^+$ T- lymphocytes coinciding with a normalization of Th1/Th2/Th3 cytokine balance. Collectively, early trials have demonstrated the efficacy and safety of MSC-transplantation for the reversal of autoimmune β-cell destruction. Further insight into the underlying mechanisms of immunomodulation by MSC is strongly warranted.

Pro-regenerative and immunomodulatory properties of MSC may also provide benefit during whole-pancreas or islet transplantation (*i.e.*, Edmonton Protocol). In 2009, Figliuzzi *et al.* demonstrated the co-implantation of BM-derived MSC increased the capillary density within islet allografts transplanted under the kidney capsule of STZ-induced diabetic rats. Rats co-injected with MSC reached normoglycemia whereas rats receiving islet allografts alone remained hyperglycemic [110]. A similar study conducted by Solari *et al.* extended these findings by demonstrating T-cells collected from MSC-treated rats produced lower levels of IFN-γ and TNF-α *in vitro* and exhibited an IL-10 secreting $CD4^+$ phenotype, suggestive of increased regulatory T- lymphocyte composition [111]. In non-human primates, Berman *et al.* demonstrated that intraportal transplantation of MHC-II mismatched allogeneic islets followed by infusion (at 4, 5, and 11 days) of donor or third party BM-MSC supported the engraftment and function of transplanted islets [95], and reversed episodes of islet graft rejection; attributed to the increase of $CD4^+$ $Foxp3^+$ regulatory T- lymphocytes *in vivo*. Although the immunomodulatory properties of MSC remain poorly understood and evidence from clinical trials is lacking, the success of future transplantation strategies involving the use of MSC to support pancreatic/islet grafts function or endogenous β-cell regeneration will rely on the ability to dampen autoimmunity in T1D.

ENDOTHELIAL PROGENITOR CELLS (EPC)

Endothelial progenitor cells (EPC), by definition, are mesodermal-derived precursor cells in the body that differentiate to produce functional endothelial cells [112]. These cells differentiate in close proximity to HPC in the developing embryo, forming "blood-islands." In 1997, Asahara *et al.* were the first group to isolate EPC from adult circulating peripheral blood and show their differentiation into endothelial cells [113]. Similar to the hematopoietic system, CD34 expression was used as the initial criteria to isolate EPC from human peripheral blood. Propagation of these circulating cells *in vitro* results in the formation of endothelial cell colony forming units (CFU-EC). These cells did not significantly incorporate into vessels *in vivo* and were controversial in regards to their origin as they shared many phenotypic markers with hematopoietic cells. Moreover, the secondary culture of these so called "endothelial precursor cells" showed progeny that were both CD45$^+$ and CD45$^-$, with colony forming assays showing a mix of EPC and HPC. A decade later, Yoder and Ingram took a different approach to isolating "true" EPC, the building blocks of blood vessels [114]. They depleted their culture of all the non-adherent CD45$^+$ cells, and the adherent population was cultured with growth factors that support endothelial cell outgrowth. In contrast to CFU-EC, these cells produced true endothelial colony forming cells (ECFC) that were purely CD45$^-$ and efficiently integrated into pre-existing and synthesized neovessels *in vivo*.

Cells that are capable of incorporating into existing vessels as well as that partake in post-natal vasculogenesis can be useful tools in regenerative medicine for the treatment of diabetes and secondary complications such as coronary and peripheral artery disease. During embryogenesis, pancreas induction occurs from the dorsal pancreatic bud arising from a portion of foregut endoderm that contacts the endothelium of the dorsal aorta and vitelline veins [115]. This concept has fuelled the concept that primitive EPC may possess intrinsic mechanisms capable of supporting endogenous β-cell mass regeneration and/or allograft survival. Additionally, in islet transplantation models and in models where β-cell injury was induced, EPC from human peripheral blood increased vascular density in either transplanted or endogenous islets respectively, providing further evidence for their use in endogenous β-cell repair [116].

Can We Utilize EPC to Support Endogenous Islet Regeneration and Revascularization?

Although research on induction of β-cell regeneration or support of islet transplantation by EPC is extremely limited, a common focus among studies is the promotion of angiogenesis within regenerating islet-like structures [117]. Islets of Langerhans are highly vascularized microstructures; however, the level of vascularization within diabetic islets is severely reduced [118]. As a result, islets experience prolonged hypoxia, facilitating ROS production and subsequent activation of apoptotic pathways [119]. Therefore, promotion of angiogenesis is essential to restore normoxia within regenerating islets. EPC respond to hypoxia-induced cytokines from ischemic tissue, allowing them to home to sites of injury. Mathews *et al.* demonstrated this concept during BM transplantation studies. They used β-cell injury induced by STZ injections and subsequently injected EPC from BM *via* the tail vein [120]. Donor as well as recipient endothelial cell frequency increased in the endocrine pancreas, reducing hypoxia at the site of islet injury. While there is still much work to be done in defining the role of EPC in islet regeneration, they accelerate revascularization in transplanted islets and provide us with a model to study how EPC can be integrated into combinatorial cellular therapies for the treatment of diabetes.

EPC-induced Vascularization Support the Survival and Function of Islet Allografts?

One of the biggest hurdles to overcome after islet transplantation in diabetic patients is the proper vascularization of donor islets within the host. Proper blood supply is so important to the function of pancreatic islets that during development, a vascular network is built surrounding individual endocrine cells prior to the formation of functional islets [121]. The regenerative properties of EPC need to be exploited for the support of islet or β-cell transplantation, as hypoxic conditions within allografts subsequently impact glucose stimulated insulin release [122]. To address this concept, Kang *et al.* injected human UCB-derived EPC into NOD/SCID mice following human islet transplantation under the kidney capsule. Co-transplantation of EPC increased the islet allograft survival and function coinciding with a reduction in measurable levels of hypoxia; resulting in increased circulating insulin leading to euglycemia by day 11 [123]. Therefore, EPC may yield the greatest benefit by the rapid induction of angiogenesis during transplantation of β-cells or islets from either postnatal or pluripotent sources.

Several groups have looked at the mobilization and recruitment of recipient EPC after the transplantation of donor islets. Since the BM has been shown to be a potent source of EPC, treatment with GM-CSF can mobilize EPC, allowing them support the revascularization of islets after implantation [116]. Brissova *et al.* were the first to suggest that harnessing both angiogenesis and vasculogenesis would be required to efficiently revascularize transplanted islets (within hours to days) to ensure high rates of survival. For example, when human islets were transplanted into immunodeficient mice, resulting vasculature surrounding the islets were surprisingly chimeric, with a mix of human intraislet and recipient mouse endothelial cells [121, 124]. Furthermore, these studies detailed the angiogenic factors that orchestrate neovascularization of pancreatic β-cells, primarily through vascular endothelial growth factor A (VEGF-A) expression by hypoxic β-cells [125, 126]. Together, these findings highlight the importance of studying recipient progenitor populations prior to transplantation, intra-islet endothelial cell populations in donor islets, and the potential for combinatorial transplantations for the treatment of diabetes mellitus.

CONCLUSION AND FUTURE DIRECTIONS

Throughout this chapter we have reviewed accumulating pre-clinical and clinical evidence demonstrating the therapeutic potential of post-natal stem cells to support the regeneration of endogenous β-cells or allogeneic islet transplantation (Fig. **2**). These emerging strategies hold great promise for effective treatment of both T1D and T2D. However, many questions surrounding the safety, feasibility, and effectiveness of postnatal cellular therapies remain to be fully addressed. Cellular therapies utilizing postnatal stem cells are currently used for the treatment of many other diseases, and therefore represent the most direct route to clinical applications for T1D or T2D.

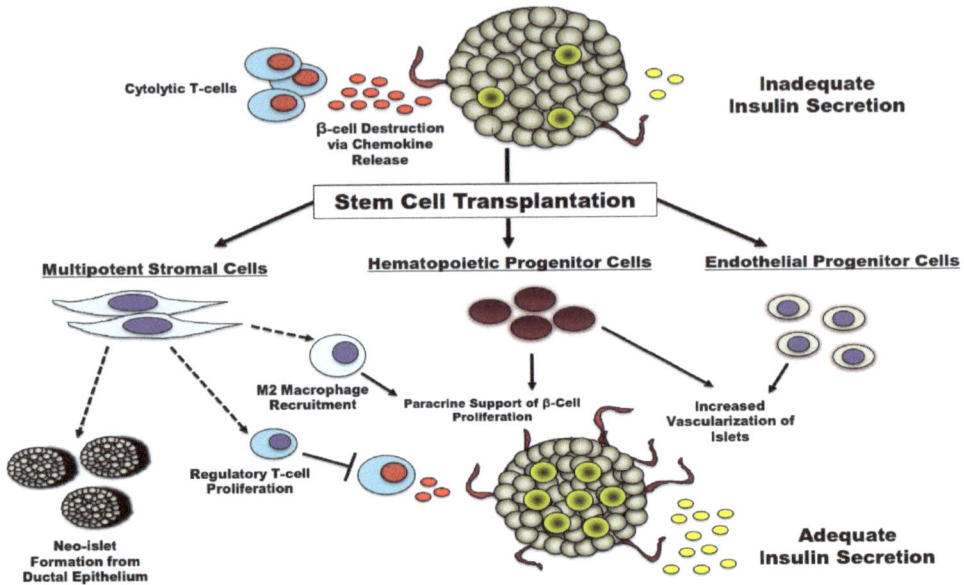

Fig. (2). Postnatal stem cells create a pro-regenerative microenvironment to increase β-cell mass and function in addition to the recruitment of regulatory T-lymphocytes. The mechanisms by which post-natal stem cells stimulate endogenous β-cell regeneration include: (1) paracrine stimulation of residual β-cell proliferation by HPC; (2) putative stimulation of new islet formation by MSC; (3) increasing vascularization *via* pro-angiogenic signalling pathways by EPC; (4) release of pro-survival and immunomodulatory chemokines to recruit pro-regenerative cell-types such as M2 macrophages or regulatory T-lymphocytes.

HPC demonstrate the ability to increase β-cell proliferation and mass, coinciding with improved glycemic levels in pre-clinical models, and HSCT has been demonstrated in early clinical trials to improve glycemic status patients with T1D. Despite this step in the right direction, further abrogation of autoimmunity will be required to effectively cure T1D. If residual β-cell mass in T1D or long-standing T2D cannot provide the adequate amount of insulin for glycemic control, combination transplantation of MSC with HPC may be developed to stimulate new islet formation and subsequent proliferation. Similarly, the potential of MSC to provide islet-regenerative and immunomodulatory cues has been repeatedly demonstrated; and is currently under investigation in multiple clinical trials. Nonetheless, many questions remain on the utilization of MSC for effective therapies such as donor source, passage number, encapsulation, and site of transplantation all require further investigation. Finally, EPC are the building blocks of new blood vessels and may be utilized to increase angiogenesis within regenerating islets or reduce hypoxia during islet transplantation. Continuing studies to identify islet-regenerative subsets by surface and/or functional marker expression, characterization of molecular signatures that stimulate islet regeneration, and specifying signal receiving cell types within the pancreas that mediate both β-cell proliferative and neogenic programs are highly warranted. Understanding which resident pancreatic cells may give rise to new β-cells in response to specific stem cell secreted effectors may permit development of cell-free regenerative therapies using a mixture of recombinant cytokines or drugs aimed at targeting β-cell regenerative programs. Therefore, elucidating the underlying mechanisms by which post-natal stem cells stimulate endogenous β-cell regeneration in human systems represent an exciting opportunity to develop curative therapies for diabetes mellitus.

CONFLICT OF INTEREST

The authors confirm that they have no conflict of interest to declare for this publication.

ACKNOWLEDGEMENTS

The authors wish to thank the Juvenile Diabetes Research Foundation (2-SR--2015-60-Q-R) and the Canadian Institutes of Health Research (MOP# 86702) for financial support of this work.

REFERENCES

[1]   Care D. Diagnosis and classification of diabetes mellitus. Diabetes Care 2014; S37-42.
[http://dx.doi.org/http://dx.doi.org/10.2337/dc14-S081]

[2]   Todd JA. Etiology of type 1 diabetes. Immunity 2010; 32(4): 457-67.
[http://dx.doi.org/10.1016/j.immuni.2010.04.001] [PMID: 20412756]

[3]   Gepts W. Pathologic anatomy of the pancreas in juvenile diabetes mellitus. Diabetes 1965; 14(10): 619-33.
[http://dx.doi.org/10.2337/diab.14.10.619] [PMID: 5318831]

[4]   International Diabetes Federation. Diabetes Atlas. 2014; pp. 1-2.

[5]   Butler AE, Janson J, Bonner-Weir S, Ritzel R, Rizza RA, Butler PC. β-cell deficit and increased β-cell apoptosis in humans with type 2 diabetes. Diabetes 2003; 52(1): 102-10.
[http://dx.doi.org/10.2337/diabetes.52.1.102] [PMID: 12502499]

[6]   Meier JJ, Lin JC, Butler AE, Galasso R, Martinez DS, Butler PC. Direct evidence of attempted β cell

regeneration in an 89-year-old patient with recent-onset type 1 diabetes. Diabetologia 2006; 49(8): 1838-44.
[http://dx.doi.org/10.1007/s00125-006-0308-2] [PMID: 16802132]

[7]   Rahier J, Guiot Y, Goebbels RM, Sempoux C, Henquin JC. Pancreatic β-cell mass in European subjects with type 2 diabetes. Diabetes Obes Metab 2008; 10 (Suppl. 4): 32-42.
[http://dx.doi.org/10.1111/j.1463-1326.2008.00969.x] [PMID: 18834431]

[8]   Guariguata L, Whiting DR, Hambleton I, Beagley J, Linnenkamp U, Shaw JE. Global estimates of diabetes prevalence for 2013 and projections for 2035. Diabetes Res Clin Pract 2014; 103(2): 137-49.
[http://dx.doi.org/10.1016/j.diabres.2013.11.002] [PMID: 24630390]

[9]   Banting FG, Best CH. The internal secretion of the pancreas. 1922. Indian J Med Res 2007; 125(3): 251-66.
[PMID: 17582843]

[10]   Walsh J, Roberts R. Pumping insulin: everything you need for success on an insulin pump. 4th ed., San Diego: Torrey Pines Press 2006.

[11]   Pelletier C, Dai S, Roberts KC, Bienek A, Onysko J, Pelletier L. Report summary. Diabetes in Canada: facts and figures from a public health perspective. Chronic Dis Inj Can 2012; 33(1): 53-4.
[PMID: 23294922]

[12]   American Diabetes Association. Economic costs of diabetes in the U.S. in 2012. Diabetes Care 2013; 36(4): 1033-46.
[http://dx.doi.org/10.2337/dc12-2625] [PMID: 23468086]

[13]   Shapiro AM, Ricordi C, Hering BJ, *et al.* International trial of the Edmonton protocol for islet transplantation. N Engl J Med 2006; 355(13): 1318-30.
[http://dx.doi.org/10.1056/NEJMoa061267] [PMID: 17005949]

[14]   Shapiro AM, Lakey JR, Ryan EA, *et al.* Islet transplantation in seven patients with type 1 diabetes mellitus using a glucocorticoid-free immunosuppressive regimen. N Engl J Med 2000; 343(4): 230-8.
[http://dx.doi.org/10.1056/NEJM200007273430401] [PMID: 10911004]

[15]   Keymeulen B, Gillard P, Mathieu C, *et al.* Correlation between β cell mass and glycemic control in type 1 diabetic recipients of islet cell graft. Proc Natl Acad Sci USA 2006; 103(46): 17444-9.
[http://dx.doi.org/10.1073/pnas.0608141103] [PMID: 17090674]

[16]   Schulz TC, Young HY, Agulnick AD, *et al.* A scalable system for production of functional pancreatic progenitors from human embryonic stem cells. PLoS One 2012; 7(5): e37004.
[http://dx.doi.org/10.1371/journal.pone.0037004] [PMID: 22623968]

[17]   Rezania A, Bruin JE, Arora P, *et al.* Reversal of diabetes with insulin-producing cells derived *in vitro* from human pluripotent stem cells. Nat Biotechnol 2014; 32(11): 1121-33.
[http://dx.doi.org/10.1038/nbt.3033] [PMID: 25211370]

[18]   Pagliuca FW, Millman JR, Gürtler M, *et al.* Generation of functional human pancreatic β cells *in vitro*. Cell 2014; 159(2): 428-39.
[http://dx.doi.org/10.1016/j.cell.2014.09.040] [PMID: 25303535]

[19]   Viacyte Regenerating Health. 2015; Available from: http://viacyte.com.

[20]   Keenan HA, Sun JK, Levine J, *et al.* Residual insulin production and pancreatic ß-cell turnover after 50 years of diabetes: Joslin Medalist Study. Diabetes 2010; 59(11): 2846-53.
[http://dx.doi.org/10.2337/db10-0676] [PMID: 20699420]

[21]   Rieck S, Kaestner KH. Expansion of β-cell mass in response to pregnancy. Trends Endocrinol Metab 2010; 21(3): 151-8.
[http://dx.doi.org/10.1016/j.tem.2009.11.001] [PMID: 20015659]

[22]   Butler AE, Cao-Minh L, Galasso R, *et al.* Adaptive changes in pancreatic β cell fractional area and β cell turnover in human pregnancy. Diabetologia 2010; 53(10): 2167-76.
[http://dx.doi.org/10.1007/s00125-010-1809-6] [PMID: 20523966]

[23]   Lingohr MK, Buettner R, Rhodes CJ. Pancreatic β-cell growth and survivala role in obesity-linked type 2 diabetes? Trends Mol Med 2002; 8(8): 375-84.
[http://dx.doi.org/10.1016/S1471-4914(02)02377-8] [PMID: 12127723]

[24]   Hess D, Li L, Martin M, *et al.* Bone marrow-derived stem cells initiate pancreatic regeneration. Nat Biotechnol 2003; 21(7): 763-70.
[http://dx.doi.org/10.1038/nbt841] [PMID: 12819790]

[25]   Ianus A, Holz GG, Theise ND, Hussain MA. *In vivo* derivation of glucose-competent pancreatic endocrine cells from bone marrow without evidence of cell fusion. J Clin Invest 2003; 111(6): 843-50.
[http://dx.doi.org/10.1172/JCI200316502] [PMID: 12639990]

[26]   Lechner A, Yang Y-G, Blacken RA, Wang L, Nolan AL, Habener JF. No evidence for significant transdifferentiation of bone marrow into pancreatic β-cells *in vivo*. Diabetes 2004; 53(3): 616-23.
[http://dx.doi.org/10.2337/diabetes.53.3.616] [PMID: 14988245]

[27]   Choi JB, Uchino H, Azuma K, *et al.* Little evidence of transdifferentiation of bone marrow-derived cells into pancreatic β cells. Diabetologia 2003; 46(10): 1366-74.
[http://dx.doi.org/10.1007/s00125-003-1182-9] [PMID: 12898006]

[28]   Kelly C, Flatt CC, McClenaghan NH. Stem cell-based approaches for the treatment of diabetes. Stem Cells Int 2011; 2011: 424986.
[http://dx.doi.org/10.4061/2011/424986] [PMID: 21716654]

[29]   Zhang C, Todorov I, Lin C-L, *et al.* Elimination of insulitis and augmentation of islet β cell regeneration *via* induction of chimerism in overtly diabetic NOD mice. Proc Natl Acad Sci USA 2007; 104(7): 2337-42.
[http://dx.doi.org/10.1073/pnas.0611101104] [PMID: 17267595]

[30]   Urbán VS, Kiss J, Kovács J, *et al.* Mesenchymal stem cells cooperate with bone marrow cells in therapy of diabetes. Stem Cells 2008; 26(1): 244-53.
[http://dx.doi.org/10.1634/stemcells.2007-0267] [PMID: 17932424]

[31]   Dor Y, Brown J, Martinez OI, Melton DA. Adult pancreatic β-cells are formed by self-duplication rather than stem-cell differentiation. Nature 2004; 429(6987): 41-6.
[http://dx.doi.org/10.1038/nature02520] [PMID: 15129273]

[32]   Nir T, Melton DA, Dor Y. Recovery from diabetes in mice by β cell regeneration. J Clin Invest 2007; 117(9): 2553-61.
[http://dx.doi.org/10.1172/JCI32959] [PMID: 17786244]

[33]   Bonner-Weir S, Baxter LA, Schuppin GT, Smith FE. A second pathway for regeneration of adult exocrine and endocrine pancreas. A possible recapitulation of embryonic development. Diabetes 1993; 42(12): 1715-20.
[http://dx.doi.org/10.2337/diab.42.12.1715] [PMID: 8243817]

[34]   Inada A, Nienaber C, Katsuta H, *et al.* Carbonic anhydrase II-positive pancreatic cells are progenitors for both endocrine and exocrine pancreas after birth. Proc Natl Acad Sci USA 2008; 105(50): 19915-9.
[http://dx.doi.org/10.1073/pnas.0805803105] [PMID: 19052237]

[35]   Xu X, DHoker J, Stangé G, *et al.* β cells can be generated from endogenous progenitors in injured adult mouse pancreas. Cell 2008; 132(2): 197-207.
[http://dx.doi.org/10.1016/j.cell.2007.12.015] [PMID: 18243096]

[36]   McCulloch EA, Till JE. The radiation sensitivity of normal mouse bone marrow cells, determined by quantitative marrow transplantation into irradiated mice. Radiat Res 1960; 13: 115-25.
[http://dx.doi.org/10.2307/3570877] [PMID: 13858509]

[37]   Spangrude GJ, Heimfeld S, Weissman IL. Purification and characterization of mouse hematopoietic stem cells. Science 1988; 241(4861): 58-62.
[http://dx.doi.org/10.1126/science.2898810] [PMID: 2898810]

[38]   Doulatov S, Notta F, Laurenti E, Dick JE. Hematopoiesis: a human perspective. Cell Stem Cell 2012; 10(2): 120-36.
[http://dx.doi.org/10.1016/j.stem.2012.01.006] [PMID: 22305562]

[39]   Kiel MJ, Yilmaz OH, Iwashita T, Yilmaz OH, Terhorst C, Morrison SJ. SLAM family receptors distinguish hematopoietic stem and progenitor cells and reveal endothelial niches for stem cells. Cell 2005; 121(7): 1109-21.
[http://dx.doi.org/10.1016/j.cell.2005.05.026] [PMID: 15989959]

[40]   Rafii S, Lyden D. Therapeutic stem and progenitor cell transplantation for organ vascularization and regeneration. Nat Med 2003; 9(6): 702-12.
[http://dx.doi.org/10.1038/nm0603-702] [PMID: 12778169]

[41]   Krause DS, Fackler MJ, Civin CI, May WS. CD34: structure, biology, and clinical utility. Blood 1996; 87(1): 1-13.
[PMID: 8547630]

[42]   Anjos-Afonso F, Currie E, Palmer HG, Foster KE, Taussig DC, Bonnet D. CD34(-) cells at the apex of the human hematopoietic stem cell hierarchy have distinctive cellular and molecular signatures. Cell Stem Cell 2013; 13(2): 161-74.
[http://dx.doi.org/10.1016/j.stem.2013.05.025] [PMID: 23910083]

[43]   Notta F, Doulatov S, Laurenti E, Poeppl A, Jurisica I, Dick JE. Isolation of single human hematopoietic stem cells capable of long-term multilineage engraftment. Science 2011; 333(6039): 218-21.
[http://dx.doi.org/10.1126/science.1201219] [PMID: 21737740]

[44]   Schatteman GC, Hanlon HD, Jiao C, Dodds SG, Christy BA. Blood-derived angioblasts accelerate blood-flow restoration in diabetic mice. J Clin Invest 2000; 106(4): 571-8.
[http://dx.doi.org/10.1172/JCI9087] [PMID: 10953032]

[45] Hasegawa Y, Ogihara T, Yamada T, *et al.* Bone marrow (BM) transplantation promotes β-cell regeneration after acute injury through BM cell mobilization. Endocrinology 2007; 148(5): 2006-15.
[http://dx.doi.org/10.1210/en.2006-1351] [PMID: 17255204]

[46] Zandstra PW, Conneally E, Petzer AL, Piret JM, Eaves CJ. Cytokine manipulation of primitive human hematopoietic cell self-renewal. Proc Natl Acad Sci USA 1997; 94(9): 4698-703.
[http://dx.doi.org/10.1073/pnas.94.9.4698] [PMID: 9114054]

[47] Csaszar E, Chen K, Caldwell J, Chan W, Zandstra PW. Real-time monitoring and control of soluble signaling factors enables enhanced progenitor cell outputs from human cord blood stem cell cultures. Biotechnol Bioeng 2013; 9: 1-7.
[PMID: 24284903]

[48] Kirouac DC, Madlambayan GJ, Yu M, Sykes EA, Ito C, Zandstra PW. Cell-cell interaction networks regulate blood stem and progenitor cell fate. Mol Syst Biol 2009; 5: 293.
[http://dx.doi.org/10.1038/msb.2009.49] [PMID: 19638974]

[49] Chute JP, Muramoto GG, Whitesides J, *et al.* Inhibition of aldehyde dehydrogenase and retinoid signaling induces the expansion of human hematopoietic stem cells. Proc Natl Acad Sci USA 2006; 103(31): 11707-12.
[http://dx.doi.org/10.1073/pnas.0603806103] [PMID: 16857736]

[50] Ghiaur G, Yegnasubramanian S, Perkins B, Gucwa JL, Gerber JM, Jones RJ. Regulation of human hematopoietic stem cell self-renewal by the microenvironments control of retinoic acid signaling. Proc Natl Acad Sci USA 2013; 110(40): 16121-6.
[http://dx.doi.org/10.1073/pnas.1305937110] [PMID: 24043786]

[51] Seitz R, Hilger A, Heiden M. Bone marrow, peripheral blood, or umbilical cord blood: does the source of allogeneic hematopoietic progenitor cells matter? J Blood Disord Transfus 2012; 01: 1-4.
[http://dx.doi.org/10.4172/2155-9864.S1-007]

[52] Storms RW, Trujillo AP, Springer JB, *et al.* Isolation of primitive human hematopoietic progenitors on the basis of aldehyde dehydrogenase activity. Proc Natl Acad Sci USA 1999; 96(16): 9118-23.
[http://dx.doi.org/10.1073/pnas.96.16.9118] [PMID: 10430905]

[53] Muramoto GG, Russell JL, Safi R, *et al.* Inhibition of aldehyde dehydrogenase expands hematopoietic stem cells with radioprotective capacity. Stem Cells 2010; 28(3): 523-34.
[PMID: 20054864]

[54] Capoccia BJ, Robson DL, Levac KD, *et al.* Revascularization of ischemic limbs after transplantation of human bone marrow cells with high aldehyde dehydrogenase activity. Blood 2009; 113(21): 5340-51.
[http://dx.doi.org/10.1182/blood-2008-04-154567] [PMID: 19324906]

[55] Putman DM, Liu KY, Broughton HC, Bell GI, Hess DA. Umbilical cord blood-derived aldehyde dehydrogenase-expressing progenitor cells promote recovery from acute ischemic injury. Stem Cells 2012; 30(10): 2248-60.
[http://dx.doi.org/10.1002/stem.1206] [PMID: 22899443]

[56] Hess DA, Meyerrose TE, Wirthlin L, *et al.* Functional characterization of highly purified human hematopoietic repopulating cells isolated according to aldehyde dehydrogenase activity. Blood 2004;

104(6): 1648-55.
[http://dx.doi.org/10.1182/blood-2004-02-0448] [PMID: 15178579]

[57]   Hess DA, Wirthlin L, Craft TP, *et al.* Selection based on CD133 and high aldehyde dehydrogenase activity isolates long-term reconstituting human hematopoietic stem cells. Blood 2006; 107(5): 2162-9.
[http://dx.doi.org/10.1182/blood-2005-06-2284] [PMID: 16269619]

[58]   Hess DA, Craft TP, Wirthlin L, *et al.* Widespread nonhematopoietic tissue distribution by transplanted human progenitor cells with high aldehyde dehydrogenase activity. Stem Cells 2008; 26(3): 611-20.
[http://dx.doi.org/10.1634/stemcells.2007-0429] [PMID: 18055447]

[59]   Bell GI, Broughton HC, Levac KD, Allan DA, Xenocostas A, Hess DA. Transplanted human bone marrow progenitor subtypes stimulate endogenous islet regeneration and revascularization. Stem Cells Dev 2012; 21(1): 97-109.
[http://dx.doi.org/10.1089/scd.2010.0583] [PMID: 21417581]

[60]   Bell GI, Meschino MT, Hughes-Large JM, Broughton HC, Xenocostas A, Hess DA. Combinatorial human progenitor cell transplantation optimizes islet regeneration through secretion of paracrine factors. Stem Cells Dev 2012; 21(11): 1863-76.
[http://dx.doi.org/10.1089/scd.2011.0634] [PMID: 22309189]

[61]   Bell GI, Putman DM, Hughes-Large JM, Hess DA. Intrapancreatic delivery of human umbilical cord blood aldehyde dehydrogenase-producing cells promotes islet regeneration. Diabetologia 2012; 55(6): 1755-60.
[http://dx.doi.org/10.1007/s00125-012-2520-6] [PMID: 22434536]

[62]   Hess DA, Hegele RA. Linking diabetes with oxidative stress, adipokines, and impaired endothelial precursor cell function. Can J Cardiol 2012; 28(6): 629-30.
[http://dx.doi.org/10.1016/j.cjca.2012.04.003] [PMID: 22575579]

[63]   Westerweel PE, Teraa M, Rafii S, *et al.* Impaired endothelial progenitor cell mobilization and dysfunctional bone marrow stroma in diabetes mellitus. PLoS One 2013; 8(3): e60357.
[http://dx.doi.org/10.1371/journal.pone.0060357] [PMID: 23555959]

[64]   Tepper OM, Galiano RD, Capla JM, *et al.* Human endothelial progenitor cells from type II diabetics exhibit impaired proliferation, adhesion, and incorporation into vascular structures. Circulation 2002; 106(22): 2781-6.
[http://dx.doi.org/10.1161/01.CIR.0000039526.42991.93] [PMID: 12451003]

[65]   Seneviratne AK, Bell GI, Cooper TT, Sherman SE, Hess DA. Islet regenerative properties of *ex vivo* expanded hematopoietic progenitor cells. Stem Cells 2016; 34: 873-87.
[http://dx.doi.org/10.1002/stem.2268] [PMID: 26676482]

[66]   Rhinn M, Dollé P. Retinoic acid signalling during development. Development 2012; 139(5): 843-58.
[http://dx.doi.org/10.1242/dev.065938] [PMID: 22318625]

[67]   Duong V, Rochette-Egly C. The molecular physiology of nuclear retinoic acid receptors. From health to disease. Biochim Biophys Acta 2011; 1812: 1023-31.

[68]   Sibley RK, Sutherland DE, Goetz F, Michael AF. Recurrent diabetes mellitus in the pancreas iso- and allograft. A light and electron microscopic and immunohistochemical analysis of four cases. Lab Invest 1985; 53(2): 132-44.

[PMID: 3894793]

[69]   Kang EM, Zickler PP, Burns S, *et al.* Hematopoietic stem cell transplantation prevents diabetes in NOD mice but does not contribute to significant islet cell regeneration once disease is established. Exp Hematol 2005; 33(6): 699-705.
[http://dx.doi.org/10.1016/j.exphem.2005.03.008] [PMID: 15911094]

[70]   Wen Y, Ouyang J, Yang R, *et al.* Reversal of new-onset type 1 diabetes in mice by syngeneic bone marrow transplantation. Biochem Biophys Res Commun 2008; 374(2): 282-7.
[http://dx.doi.org/10.1016/j.bbrc.2008.07.016] [PMID: 18625200]

[71]   Voltarelli JC, Couri CE, Stracieri AB, *et al.* Autologous nonmyeloablative hematopoietic stem cell transplantation in newly diagnosed type 1 diabetes mellitus. JAMA 2007; 297(14): 1568-76.
[http://dx.doi.org/10.1001/jama.297.14.1568] [PMID: 17426276]

[72]   Couri CE, Oliveira MC, Stracieri AB, *et al.* C-peptide levels and insulin independence following autologous nonmyeloablative hematopoietic stem cell transplantation in newly diagnosed type 1 diabetes mellitus. JAMA 2009; 301(15): 1573-9.
[http://dx.doi.org/10.1001/jama.2009.470] [PMID: 19366777]

[73]   Haller MJ, Viener HL, Wasserfall C, Brusko T, Atkinson MA, Schatz DA. Autologous umbilical cord blood infusion for type 1 diabetes. Exp Hematol 2008; 36(6): 710-5.
[http://dx.doi.org/10.1016/j.exphem.2008.01.009] [PMID: 18358588]

[74]   Friedenstein AJ, Chailakhjan RK, Lalykina KS. The development of fibroblast colonies in monolayer cultures of guinea-pig bone marrow and spleen cells. Cell Tissue Kinet 1970; 3(4): 393-403.
[PMID: 5523063]

[75]   Caplan AI. Mesenchymal stem cells. J Orthop Res 1991; 9(5): 641-50.
[http://dx.doi.org/10.1002/jor.1100090504] [PMID: 1870029]

[76]   Bianco P, Robey PG, Saggio I, Riminucci M. Mesenchymal stem cells in human bone marrow (skeletal stem cells): a critical discussion of their nature, identity, and significance in incurable skeletal disease. Hum Gene Ther 2010; 21(9): 1057-66.
[http://dx.doi.org/10.1089/hum.2010.136] [PMID: 20649485]

[77]   da Silva Meirelles L, Caplan AI, Nardi NB. In search of the *in vivo* identity of mesenchymal stem cells. Stem Cells 2008; 26(9): 2287-99.
[http://dx.doi.org/10.1634/stemcells.2007-1122] [PMID: 18566331]

[78]   Dominici M, Le Blanc K, Mueller I, *et al.* Minimal criteria for defining multipotent mesenchymal stromal cells. The International Society for Cellular Therapy position statement. Cytotherapy 2006; 8(4): 315-7.
[http://dx.doi.org/10.1080/14653240600855905] [PMID: 16923606]

[79]   Crisan M, Yap S, Casteilla L, *et al.* A perivascular origin for mesenchymal stem cells in multiple human organs. Cell Stem Cell 2008; 3(3): 301-13.
[http://dx.doi.org/10.1016/j.stem.2008.07.003] [PMID: 18786417]

[80]   Méndez-Ferrer S, Michurina TV, Ferraro F, *et al.* Mesenchymal and haematopoietic stem cells form a unique bone marrow niche. Nature 2010; 466(7308): 829-34.
[http://dx.doi.org/10.1038/nature09262] [PMID: 20703299]

[81]   Bianco P, Cao X, Frenette PS, *et al.* The meaning, the sense and the significance: translating the science of mesenchymal stem cells into medicine. Nat Med 2013; 19(1): 35-42.
[http://dx.doi.org/10.1038/nm.3028] [PMID: 23296015]

[82]   Deans RJ, Moseley AB. Mesenchymal stem cells: biology and potential clinical uses. Exp Hematol 2000; 28(8): 875-84.
[http://dx.doi.org/10.1016/S0301-472X(00)00482-3] [PMID: 10989188]

[83]   Chen L, Jiang X, Yang L. Differentiation of rat marrow mesenchymal stem cells into pancreatic islet β-cells. World J Gastroenterol 2004; 10: 3016-20.
[http://dx.doi.org/10.3748/wjg.v10.i20.3016]

[84]   Abdi R, Fiorina P, Adra CN, Atkinson M, Sayegh MH. Immunomodulation by mesenchymal stem cells: a potential therapeutic strategy for type 1 diabetes. Diabetes 2008; 57(7): 1759-67.
[http://dx.doi.org/10.2337/db08-0180] [PMID: 18586907]

[85]   Lee RH, Seo MJ, Reger RL, *et al.* Multipotent stromal cells from human marrow home to and promote repair of pancreatic islets and renal glomeruli in diabetic NOD/scid mice. Proc Natl Acad Sci USA 2006; 103(46): 17438-43.
[http://dx.doi.org/10.1073/pnas.0608249103] [PMID: 17088535]

[86]   Ezquer M. Mesenchymal stem cell therapy in type 1 diabetes mellitus and its main complications: from experimental findings to clinical practice. J Stem Cell Res Ther 2014; 04: 1-15.
[http://dx.doi.org/10.4172/2157-7633.1000227]

[87]   Sordi V, Malosio ML, Marchesi F, *et al.* Bone marrow mesenchymal stem cells express a restricted set of functionally active chemokine receptors capable of promoting migration to pancreatic islets. Blood 2005; 106(2): 419-27.
[http://dx.doi.org/10.1182/blood-2004-09-3507] [PMID: 15784733]

[88]   Cheng H, Zhang YC, Wolfe S, *et al.* Combinatorial treatment of bone marrow stem cells and stromal cell-derived factor 1 improves glycemia and insulin production in diabetic mice. Mol Cell Endocrinol 2011; 345(1-2): 88-96.
[http://dx.doi.org/10.1016/j.mce.2011.07.024] [PMID: 21801807]

[89]   Lee RH, Pulin AA, Seo MJ, *et al.* Intravenous hMSCs improve myocardial infarction in mice because cells embolized in lung are activated to secrete the anti-inflammatory protein TSG-6. Cell Stem Cell 2009; 5(1): 54-63.
[http://dx.doi.org/10.1016/j.stem.2009.05.003] [PMID: 19570514]

[90]   Wang MK, Sun HQ, Xiang YC, Jiang F, Su YP, Zou ZM. Different roles of TGF-β in the multi-lineage differentiation of stem cells. World J Stem Cells 2012; 4(5): 28-34.
[http://dx.doi.org/10.4252/wjsc.v4.i5.28] [PMID: 22993659]

[91]   Gao X, Song L, Shen K, Wang H, Niu W, Qin X. Transplantation of bone marrow derived cells promotes pancreatic islet repair in diabetic mice. Biochem Biophys Res Commun 2008; 371(1): 132-7.
[http://dx.doi.org/10.1016/j.bbrc.2008.04.033] [PMID: 18420028]

[92]   Si Y, Zhao Y, Hao H, *et al.* Infusion of mesenchymal stem cells ameliorates hyperglycemia in type 2 diabetic rats: identification of a novel role in improving insulin sensitivity. Diabetes 2012; 61(6): 1616-25.

[http://dx.doi.org/10.2337/db11-1141] [PMID: 22618776]

[93]   Kern S, Eichler H, Stoeve J, Klüter H, Bieback K. Comparative analysis of mesenchymal stem cells from bone marrow, umbilical cord blood, or adipose tissue. Stem Cells 2006; 24(5): 1294-301.
[http://dx.doi.org/10.1634/stemcells.2005-0342] [PMID: 16410387]

[94]   François M, Romieu-Mourez R, Li M, Galipeau J. Human MSC suppression correlates with cytokine induction of indoleamine 2,3-dioxygenase and bystander M2 macrophage differentiation. Mol Ther 2012; 20(1): 187-95.
[http://dx.doi.org/10.1038/mt.2011.189] [PMID: 21934657]

[95]   Berman DM, Willman MA, Han D, *et al.* Mesenchymal stem cells enhance allogeneic islet engraftment in nonhuman primates. Diabetes 2010; 59(10): 2558-68.
[http://dx.doi.org/10.2337/db10-0136] [PMID: 20622174]

[96]   Hu J, Yu X, Wang Z, *et al.* Long term effects of the implantation of Whartons jelly-derived mesenchymal stem cells from the umbilical cord for newly-onset type 1 diabetes mellitus. Endocr J 2013; 60(3): 347-57.
[http://dx.doi.org/10.1507/endocrj.EJ12-0343] [PMID: 23154532]

[97]   Jiang R, Han Z, Zhuo G, *et al.* Transplantation of placenta-derived mesenchymal stem cells in type 2 diabetes: a pilot study. Front Med 2011; 5(1): 94-100.
[http://dx.doi.org/10.1007/s11684-011-0116-z] [PMID: 21681681]

[98]   Li FX, Zhu JW, Tessem JS, *et al.* The development of diabetes in E2f1/E2f2 mutant mice reveals important roles for bone marrow-derived cells in preventing islet cell loss. Proc Natl Acad Sci USA 2003; 100(22): 12935-40.
[http://dx.doi.org/10.1073/pnas.2231861100] [PMID: 14566047]

[99]   Tessem JS, Jensen JN, Pelli H, *et al.* Critical roles for macrophages in islet angiogenesis and maintenance during pancreatic degeneration. Diabetes 2008; 57(6): 1605-17.
[http://dx.doi.org/10.2337/db07-1577] [PMID: 18375440]

[100] Xiao X, Gaffar I, Guo P, *et al.* M2 macrophages promote β-cell proliferation by up-regulation of SMAD7. Proc Natl Acad Sci USA 2014; 111(13): E1211-20.
[http://dx.doi.org/10.1073/pnas.1321347111] [PMID: 24639504]

[101] Cao X, Han ZB, Zhao H, Liu Q. Transplantation of mesenchymal stem cells recruits trophic macrophages to induce pancreatic β cell regeneration in diabetic mice. Int J Biochem Cell Biol 2014; 53: 372-9.
[http://dx.doi.org/10.1016/j.biocel.2014.06.003] [PMID: 24915493]

[102] Zhao Y, Jiang Z, Zhao T, *et al.* Reversal of type 1 diabetes *via* islet β cell regeneration following immune modulation by cord blood-derived multipotent stem cells. BMC Med 2012; 10: 3.
[http://dx.doi.org/10.1186/1741-7015-10-3] [PMID: 22233865]

[103] Takiishi T, Korf H, Van Belle TL, *et al.* Reversal of autoimmune diabetes by restoration of antigen-specific tolerance using genetically modified Lactococcus lactis in mice. J Clin Invest 2012; 122(5): 1717-25.
[http://dx.doi.org/10.1172/JCI60530] [PMID: 22484814]

[104] Fiorina P, Jurewicz M, Augello A, *et al.* Immunomodulatory function of bone marrow-derived

mesenchymal stem cells in experimental autoimmune type 1 diabetes. J Immunol 2009; 183(2): 993-1004.
[http://dx.doi.org/10.4049/jimmunol.0900803] [PMID: 19561093]

[105] Engela AU, Baan CC, Dor FJ, Weimar W, Hoogduijn MJ. On the interactions between mesenchymal stem cells and regulatory T cells for immunomodulation in transplantation. Front Immunol 2012; 3: 126.
[http://dx.doi.org/10.3389/fimmu.2012.00126] [PMID: 22629256]

[106] Domínguez-Bendala J, Lanzoni G, Inverardi L, Ricordi C. Concise review: mesenchymal stem cells for diabetes. Stem Cells Transl Med 2012; 1(1): 59-63.
[http://dx.doi.org/10.5966/sctm.2011-0017] [PMID: 23197641]

[107] Hof-Nahor I, Leshansky L, Shivtiel S, *et al.* Human mesenchymal stem cells shift CD8+ T cells towards a suppressive phenotype by inducing tolerogenic monocytes. J Cell Sci 2012; 125(Pt 19): 4640-50.
[http://dx.doi.org/10.1242/jcs.108860] [PMID: 22767507]

[108] Madec AM, Mallone R, Afonso G, *et al.* Mesenchymal stem cells protect NOD mice from diabetes by inducing regulatory T cells. Diabetologia 2009; 52(7): 1391-9.
[http://dx.doi.org/10.1007/s00125-009-1374-z] [PMID: 19421731]

[109] Boumaza I, Srinivasan S, Witt WT, *et al.* Autologous bone marrow-derived rat mesenchymal stem cells promote PDX-1 and insulin expression in the islets, alter T cell cytokine pattern and preserve regulatory T cells in the periphery and induce sustained normoglycemia. J Autoimmun 2009; 32(1): 33-42.
[http://dx.doi.org/10.1016/j.jaut.2008.10.004] [PMID: 19062254]

[110] Figliuzzi M, Cornolti R, Perico N, *et al.* Bone marrow-derived mesenchymal stem cells improve islet graft function in diabetic rats. Transplant Proc 2009; 41(5): 1797-800.
[http://dx.doi.org/10.1016/j.transproceed.2008.11.015] [PMID: 19545731]

[111] Solari MG, Srinivasan S, Boumaza I, *et al.* Marginal mass islet transplantation with autologous mesenchymal stem cells promotes long-term islet allograft survival and sustained normoglycemia. J Autoimmun 2009; 32(2): 116-24.
[http://dx.doi.org/10.1016/j.jaut.2009.01.003] [PMID: 19217258]

[112] Risau W. Mechanisms of angiogenesis. Nature 1997; 386(6626): 671-4.
[http://dx.doi.org/10.1038/386671a0] [PMID: 9109485]

[113] Asahara T, Murohara T, Sullivan A, *et al.* Isolation of putative progenitor endothelial cells for angiogenesis. Science 1997; 275(5302): 964-7.
[http://dx.doi.org/10.1126/science.275.5302.964] [PMID: 9020076]

[114] Yoder MC, Mead LE, Prater D, *et al.* Plenary paper redefining endothelial progenitor cells *via* clonal analysis and hematopoietic stem/progenitor cell principals. Blood 2007; 109: 1-3.
[http://dx.doi.org/10.1182/blood-2006-08-043471]

[115] Lammert E, Cleaver O, Melton D. Induction of pancreatic differentiation by signals from blood vessels. Science 2001; 294(5542): 564-7.
[http://dx.doi.org/10.1126/science.1064344] [PMID: 11577200]

[116] Contreras JL, Smyth CA, Eckstein C, *et al.* Peripheral mobilization of recipient bone marrow-derived endothelial progenitor cells enhances pancreatic islet revascularization and engraftment after intraportal transplantation. Surgery 2003; 134(2): 390-8.
[http://dx.doi.org/10.1067/msy.2003.250] [PMID: 12947346]

[117] Quaranta P, Antonini S, Spiga S, *et al.* Co-transplantation of endothelial progenitor cells and pancreatic islets to induce long-lasting normoglycemia in streptozotocin-treated diabetic rats. PLoS One 2014; 9(4): e94783.
[http://dx.doi.org/10.1371/journal.pone.0094783] [PMID: 24733186]

[118] Denis MC, Mahmood U, Benoist C, Mathis D, Weissleder R. Imaging inflammation of the pancreatic islets in type 1 diabetes. Proc Natl Acad Sci USA 2004; 101(34): 12634-9.
[http://dx.doi.org/10.1073/pnas.0404307101] [PMID: 15304647]

[119] Bento CF, Pereira P. Regulation of hypoxia-inducible factor 1 and the loss of the cellular response to hypoxia in diabetes. Diabetologia 2011; 54(8): 1946-56.
[http://dx.doi.org/10.1007/s00125-011-2191-8] [PMID: 21614571]

[120] Mathews V, Hanson PT, Ford E, Fujita J, Polonsky KS, Graubert TA. Recruitment of bone marrow-derived endothelial cells to sites of pancreatic β-cell injury. Diabetes 2004; 53(1): 91-8.
[http://dx.doi.org/10.2337/diabetes.53.1.91] [PMID: 14693702]

[121] Brissova M, Powers AC. Revascularization of transplanted islets: can it be improved? Diabetes 2008; 57(9): 2269-71.
[http://dx.doi.org/10.2337/db08-0814] [PMID: 18753672]

[122] Jansson L, Carlsson PO. Graft vascular function after transplantation of pancreatic islets. Diabetologia 2002; 45(6): 749-63.
[http://dx.doi.org/10.1007/s00125-002-0827-4] [PMID: 12107718]

[123] Kang S, Park HS, Jo A, *et al.* Endothelial progenitor cell cotransplantation enhances islet engraftment by rapid revascularization. Diabetes 2012; 61(4): 866-76.
[http://dx.doi.org/10.2337/db10-1492] [PMID: 22362173]

[124] Nyqvist D, Köhler M, Wahlstedt H, Berggren PO. Donor islet endothelial cells participate in formation of functional vessels within pancreatic islet grafts. Diabetes 2005; 54(8): 2287-93.
[http://dx.doi.org/10.2337/diabetes.54.8.2287] [PMID: 16046293]

[125] Brissova M, Shostak A, Shiota M, *et al.* Pancreatic islet production of vascular endothelial growth factora is essential for islet vascularization, revascularization, and function. Diabetes 2006; 55(11): 2974-85.
[http://dx.doi.org/10.2337/db06-0690] [PMID: 17065333]

[126] Brissova M, Aamodt K, Brahmachary P, *et al.* Islet microenvironment, modulated by vascular endothelial growth factor-A signaling, promotes β cell regeneration. Cell Metab 2014; 19(3): 498-511.
[http://dx.doi.org/10.1016/j.cmet.2014.02.001] [PMID: 24561261]

SUBJECT INDEX

A

Acetate 115, 116
Acetylation 68, 69
Acetylcholine 95, 96, 99, 110, 116
Acinar cells 5
Activity, methyltransferase 30
Adipocytes 201, 202
Adiponectin 127
Aging 20, 24, 25, 29, 33, 58, 70, 71, 73
  cellular 58, 70, 71, 73
  effects of 20, 24, 25, 29, 33
ALDHhi cells 197, 198, 199, 203
  fresh UCB 199
  human 198
  isolated 199
  iv-injected 198
ALDHhi UCB cells 199
Allogeneic HSCT 200, 201
Allogeneic islets 190, 208
  mismatched 208
Allogeneic transplantation 190, 207
Alterations in islets mass 87
Amino acids 7, 72, 107, 109
Anorexigenic neuropeptides 97, 98
Anti-diabetogenic effects 128
Architecture 46, 84
  altered islet 84
  human islet 46
Artificial rearing technique 86, 87
ATP, cytosolic 91, 92
Autocrine signaling 112, 113, 115
Autoimmune diabetes 48, 148
  adult-onset 48
Autologous BM cells 199
Autologous transplantation 190

B

B-cell ablation 25, 28, 197
B-cell apoptosis 21, 22
  glucose-induced 21
B-cell death, pancreatic 191
B-cell dedifferentiation 176, 179

B-cell destruction, autoimmune 207, 208
B-cell development 3, 7, 8, 9, 10, 11, 13, 69
B-cell differentiation 8, 9, 10
B-cell Function 11, 21, 49, 174, 204
  abnormal 11
  declining 21
  increased 21, 204
  reduced 49
  retained 174
B-cell growth 20, 24, 26, 28, 29, 58
  adaptive 24, 26
B-cell injury 209, 210
B-cell/islet density 44
B-cell loss 48
B-cell lymphoma 144
B-cell mass 7, 21, 30, 33, 42, 48, 60, 64, 70, 72, 169, 179, 191, 198, 207
  increased 21, 30, 33
  increased pancreatic 64
  loss of 179, 191
  pancreatic 21, 42
  recovery of 60, 198
  reduced 7, 21, 48, 70, 191
  reductions in 70, 72
  restoration of 169, 207
B-cell mass quantification 42, 45
B cell neogenesis 85, 108, 126, 143
B-cell overexpression 29, 30
B-cell pathophysiology 42
B-cell phenotypes 69, 72, 178
B-cell plasticity 64, 67
B-cell progenitor cells 61
B-cell proliferation 22, 23, 24, 25, 26, 27, 28, 29, 30, 31, 32, 64, 195, 199, 212
  adaptive 23, 25, 28
  age-restricted 29
  basal 26
  compensatory 32
  decreased 28, 29, 30, 31
  glucose-induced 27, 28, 29
  human 22, 26
  increased 26
  increasing 25, 32
  induced 27, 28, 29
  inhibited 64

R

Receptor activity-modifying proteins (RAMP) 114
Receptor expression 26, 27
Receptors 59, 114, 140
    calcitonin 114
    lacking pancreatic HGF 140
    tissue glucocorticoid 59
Reduction, progressive 69
Regenerating β-cells 204
Regenerating islets 196, 197, 198, 210, 213
    surrounded 198
Regeneration 20, 25, 29, 32, 169, 174, 176, 177, 180, 190, 207, 209, 211
    endogenous β-cell mass 209
Regenerative capacity 24, 25, 180, 194, 199, 203, 205
Regenerative functions 190, 199, 200
Regenerative niche for new islet formation 203
Regenerative therapies 20, 197, 198, 202
Regions, gene promoter 68
Regulation, epigenetic 20, 27, 30, 31
Resident stem/progenitor cells 177
Restriction, intrauterine growth 58, 71, 72
Retinoic acid (RA) 172, 173, 197, 199
Reverse diabetogenic effects 128
Reverse hyperglycemia 199, 200
Risk, postnatal diabetes 70

S

Short chain fatty acids (SCFAs) 115, 116
Signals, mitogenic 20, 26, 28, 32
Sirt1-mediated glucose stimulate insulin release 21
Somatolactogenic hormones 123, 127, 129, 131, 132, 133, 135, 137, 150
Sources, stem cell 171, 173, 180
Stem cells 3, 12, 169, 171, 173, 194, 195, 197, 201, 211
    embryonic 12, 171, 173
    human induced pluripotent 169
    induced pluripotent 3
    marrow stromal 201
    mesenchymal 194, 201
    neural 171

pancreatic 178
post-natal 194, 195, 211
purify 197
putative resident 169
Stem cells biology 9
Stimulate islet regeneration 213
Stimulus, important physiological 91
Streptozotocin diabetes 139
Stress, physiological 71, 149
STZ-induced diabetic rats 208
Supplementation, leucine 7, 8
Support β-cell regeneration 206, 207

T

Tagged non-β-cells 179
T-cells 208
Telomere length 71
Temporal diabetes 128
Theophylline 138
Thymidine 22, 23
Tissues 46, 47, 61, 70, 107, 108, 109, 129, 191, 192
    donor pancreas 192
    endocrine 46, 47
    non-pancreatic 61
    pancreatic 129
    peripheral 70, 107, 108, 109, 191
T-lymphocytes, regulatory 200, 201, 207, 212
Toxins, environmental 58, 70
Transcription factors 3, 4, 8, 32, 68, 69, 151
Transdifferentiation 176, 194
Trans-differentiation 175
Transplantation, direct intra-pancreatic 198
Transplantation of freshly-isolated BM ALDHhi cells 198
Transverse mesocolon 43
Trefoil factors (TFFs) 147
Trimethylates 31
Trimethylation 30, 31
TrxG proteins 31, 33
Tryptophan 145
TSC2 gene in β-cells 64
Tuberous sclerosis 64, 65
Tumors, pancreatic 45

U

www.ingramcontent.com/pod-product-compliance
Lightning Source LLC
Chambersburg PA
CBHW080019240326
41598CB00075B/320

* 9 7 8 1 6 8 1 0 8 3 6 6 7 *